STATISTICS FOR THE HUMAN SCIENCES

TO THE MEMORY OF
OLIVER DAVID HAMMERTON
1962–1973

Statistics for the human sciences

M. Hammerton

Longman
London and New York

Longman Group Limited
Burnt Mill
Harlow
Essex CM20 2JE

Published in the United States of America by Longman Inc., New York

Associated companies, branches and representatives
throughout the world

First published 1975

Library of Congress Cataloging in Publication Data
Hammerton, M
    Statistics for the human sciences.

    Bibliography: p.
    Includes index.
    1. Statistics. I. Title.
HA29.H2437 519.5 75—5873
ISBN 0 582 44273 7 pbk.

Set in IBM Journal, 10 on 12pt.

Printed in Great Britain by
Whitstable Litho Ltd

Contents

Acknowledgements

We are grateful to the following for permission to reproduce copyright material:

American Statistical Association for a table adapted from Kruskal & Wallis, *Journal of American Statistical Association*, 47, 1952 and a table adapted from Friedman, *Journal of American Statistical Association*, 32, 1937; Bulletin Institute of Educational Research for two tables adapted from Auble, *Bulletin Institute Educational Research, Indiana University*, 1953, 1, No. 2; Charles Griffin & Company Ltd for a table adapted from Kendall *Rank Correlation Methods*, 1948 (4th edition 1970), reproduced by permission of the Publishers Charles Griffin & Company Ltd., London and High Wycombe; The Institute of Mathematical Statistics for a table adapted from Milton Friedman, March 1940 "A comparison of alternative tests of significance for the problems of m rankings" from *Annals of Mathematical Statistics*, Vol. XI 1940, a table adapted from E. G. Olds "Distributions of sums of squares of rank differences for small numbers of individuals" and "The 5% significance levels for sums of squares of rank differences and a correction" from *Annals of Mathematical Statistics*, Vol. IX, 1938 and Vol. XX, 1949, and two tables adapted from Frieda S. Swed and C. Eisenhart "Tables for testing randomness of grouping in a sequence of alternatives" from *Annals of Mathematical Statistics*, Vol. XIV, 1943; Iowa State University Press for a table adapted from *Statistical Methods* by George W. Snedecor and William G. Cochran, sixth edition. © 1967 by Iowa State University Press, Ames, Iowa; Lederle Laboratories, A Division of American Cyanamid Company for a table adapted from *Some Rapid Approximate Statistical Procedures* by Wilcoxon and Wilcox, and Longman Group Ltd for four tables (one modified) from Fisher and Yates: *Statistical Tables for Biological, Agricultural and Medical Research*, published by Longman Group Ltd, London (previously published by Oliver & Boyd, Edinburgh), and by permission of the authors and publishers.

We regret we have been unable to trace the copyright holder of the

following material and would appreciate receiving any information that
would enable us to do so:

A table which has been adapted from p. 458 of *Statistical Inference* by
Walker & Lev, 1953.

Introductory note

About this book and its readers

When deciding whether a particular book — such as this one — is the one you want, there are two questions which you have to answer: what is in it? and for whom is it written? Oddly enough, these are exactly the questions which an author has to keep in mind while he is writing. Now, unfortunately, publishers' blurbs have inflated themselves out of the market: even the best of them are liable to tell you that *this* book provides *the* answers precisely for *you*. It therefore seems a good idea that I should begin by briefly explaining to whom I intend to convey what, and how I have tried to do it.

First then, who are you? You are a person of either sex and almost any age who is interested in one or all of the disciplines concerned with human beings, such as psychology, social studies, history or economics. Your mathematical equipment is equivalent to O-level; and you wish to acquaint yourself with some of the statistical ideas and techniques useful in your area of interest. You are pretty bright, and are prepared to put in a certain amount of serious effort.

This book is intended to enable you to fulfil that wish. You will find a general account of some of the basic notions of statistics, and a more detailed account of the ideas underlying, and the mode of employing some of the tests and measures popular in your area. This is not intended to be a cook-book. Cook-books are useful when you already know the ideas behind and the limitations of the recipes; and I suppose that you do not know them. I have tried to provide sufficient background information on each technique discussed to enable you to use it sensibly and appropriately.

There are a great many statistical techniques; and any book which attempted to expound all of them would be impossibly bulky. Therefore, there is a problem of selection. I have, for example, not included an exposition of regression, or of the Q-test. The principle of inclusion I have mainly used is frequency of appearance in a variety of journals.

Having decided upon what I wanted to do, there arose the question of how to do it. I wish to be clear; but what seems clear to me might be far more obscure than I think. I wish to be accurate; but I am not a professional mathematician; and it is not impossible that I might mislead inadvertently. I therefore adopted the following procedure: I obtained the cooperation of a highly intelligent arts graduate whose mathematical training amounted to O-level twenty years ago, and of a professional statistician. Each section was re-written until the former found it lucid, and the latter approved it as accurate. Thus, in so far as I have succeeded, much credit must go to them; but any failings are mine alone.

If you now think that this sounds the book for you, I hope you will not be disappointed.

I would like to thank the helpers I mentioned: my wife Elizabeth Hammerton, and my friend Dr P. M. E. Altham, and also my secretary Mrs E. Larby who prepared the typescript for press and who can read even my handwriting.

1

About statistics

Chemists are a fortunate race of mortals. The objects of their study display a remarkable uniformity in both space and time, and retain that uniformity through the most elaborate transformations. A sample of chlorine will have the same properties today as it had last week, and the same as another sample anywhere else in the world. It will always combine in exactly the same proportions with other substances; and, if separated out after each combination, it will recombine in the same way as before, while exhibiting no change in its other properties.

The scientist who studies human beings, on the other hand, enjoys no such advantages. A human being under study (in technical parlance: a "subject") is not the same as he was last week; he has learned some things and forgotten others. Neither is a subject drawn from one human group identical with all others from his group: he may be more or less intelligent, more or less acute in hearing, more or less endowed in a thousand ways. Nor is a subject from one group identical with one from another group: there may be genetic differences; and there are certainly social ones whose subtle effects as yet defy clear analysis. Above all, perhaps, from the point of view of the experimenter, a test cannot be repeated upon a subject with any expectation of unchanged results: the oldest dogs will learn new tricks; and a human being who has experienced a particular test will approach it a second time with a host of expectations and strategies which he did not have when it was new to him.

Faced with such an alarming array of problems, it is not very surprising that many persons retreat from any attempt to achieve scientific under-standing of the ways of mankind. Such a retreat may take the common-nonsense view that we understand "human nature" already by some inner light; or it may take the opposite line that no understanding of any kind is possible; or (since few can accept ignorance with equanimity) it may involve a claim to a pseudo-scientific understanding, exemplified by the amusing fantasies of psycho-analysis or the much less amusing dogmas of other fashionable ideologies. Nevertheless, such ignominious responses are not unavoidable or universal. The way ahead for scientific progress in the human sciences will surely prove difficult; but there is no reason to believe that it will prove impossible.

Fortunately, there are sound experimental techniques for coping with the great variability which exists both between subjects, and within each subject over time. Also, there are sound techniques for summarizing and evaluating the results so obtained. This book will be chiefly concerned with the second problem: the summarizing and evaluation of results; but it will at times be necessary to deal with the first problem, as well as to range into wider fields.

The chief technique for dealing with data which are characterized by great variability is called "statistics". Unfortunately, the misrepresentation of data by interested parties in order to deceive the innumerate is so widespread that the method has acquired a bad name. To say of some finding that it is "only statistical" is to damn it in some quarters: the phrase carries an insinuation of unsoundness, even of disreputability. Yet this is no wiser than to condemn a proposition because it is "only stated in words" on the grounds that words are often used to deceive. Properly used and understood, statistics are a vital tool for understanding and a stout shield against deceivers.

A popular stalling technique in debate is to demand a definition. "What is your definition of democracy/justice/honesty/what-have-you?", will frequently stop an amateur debater in his tracks; but let us remember that it is *only* a stalling technique. We can make a lot of progress — indeed we generally do — before, or even without, sharp definitions. I doubt very much whether I could produce a really watertight and invariably applicable definition of — say — "war"; yet we all know, in a broad way, what we mean by it. If the BBC News Bulletin should ever tell us that "war" has broken out between nations A and B, we would not dash to our dictionaries to clear our minds about what had happened. Similarly, we can make a little progress in our understanding of statistics before we essay the precise definition of our terms.

Let us, then, first consider a statistical statement which we all accept and understand, and whose limitations we more or less intuitively grasp. The statement we shall start with is: "Men are taller than women."

This is a statement which satisfies our requirements for the moment; and it is worth pursuing in some detail. For example, consider my friends Mrs X, who stands 175 cm on her bare feet, and Mr Y, who similarly stands 170 cm. Do they disprove the statement? Not at all; it is not denied that there are tall women and short men; the statement is taken to imply that the average (a word we will come back to) of women's heights is less than of men's.

Here then is our first important point. 1: *Statistical statements are about populations, not about individuals.* (In statistical parlance, the totality of individuals being considered is always called the "population". It is not necessarily a population of human beings, or of animals; it might be bars of chocolate, galaxies, bullets, or anything we care to discuss.)

Now how do we know that our statement is true? By measuring people's heights, to be sure. But supposing we had no rulers, would we then be condemned to ignorance? We would not; for we could take a population, line the members up in order of height — which can be *compared*, though not *measured* without scales — and see where the balance of men and women lay. This point is important, for it is often possible to rank people for characteristics even when precise measurement is impossible.

Here, then, are two more points: 2: *Statistical statements, though concerned with populations, are derived from individual observations.* 3: *Absolute measurements are not essential for validity as long as individuals can be unambiguously ranked for the characteristic under consideration.*

In practice, of course, it would be impossible to measure or line up all adult human beings to make the comparisons we have mentioned; and no man has actually seen more than a small fraction of that population. This, quite rightly, does not trouble us and we are satisfied with our sample surveys, even though they may (as in this case) be pretty rough ones. But supposing we brought a group of Swedish women, whose average height is about 165 cm, and compared them with a group of Congolese Pigmy men, whose average height is around 150 cm, would you accept this as evidence against our statement? Of course not. You would, quite rightly, point out that these samples of men and women were taken from noticeably different sub-groups of the human population, and that, to be just, the samples should be directly comparable.

Thus we have two more points to keep in mind: 4: *Statistical statements can be validly inferred from sample studies*; but (and it is a big "but") 5: *Care must be taken in selecting and ensuring the comparability of samples.* We shall see in a later chapter that sampling technique is far from easy, and beset with knotty problems.

Still following our example, let us suppose that an intrepid space-traveller from Tau Ceti III (it is amazing how often that particular planet turns up in works of science fiction) collected a truly random sample (we shall have to discuss this concept in more detail later) of 100 men and 100 women, measured their heights, and inferred our statement from his data. We could ask him whether he was quite sure of its truth, and he might reply: "I'm not absolutely sure, of course; but I'm prepared to offer a bet on it." Pressing the point, we could ask him what he would consider to be reasonable odds; and this would be a very important question indeed. Being a competent statistician (all the Cetians are) his reply might be something like: "The probability of my finding arising out of pure chance is less than 1 in 1000, but more than 1 in 10,000."

Here we have, for the first time, gone beyond the scope of general common sense, and entered the realms of technical statistics. Our Cetian friend has stated a probability; and we are thus compelled to face certain quite fundamental questions, which we must try to answer before we proceed. Statistics is a useful art, whose tools are derived from a branch of study known as the theory of probability. This theory is one of the most subtle and tricky in mathematics. Fortunately, we will not need to go far into its formidable technicalities; but we must grasp some of the basic concepts, especially what do we mean by and how do we measure "Probability".

We use the word in a variety of contexts: "It will probably rain tomorrow"; "In the light of the evidence, ESP phenomena are probably spurious"; "The probability of heads is half"; "John will probably get a first". All these statements have at least this in common: they all refer to matters which are *uncertain*. Unfortunately, it is not agreed among experts whether the technical meaning of the term can or cannot be used in all these contexts, there being two main interpretations in the field at the moment. These are the Frequentist on the one hand and the Bayesian on the other. (The name Bayesian derives from the Revd Thomas Bayes, an English clergyman and mathematician, who published (1763) an important paper on probability, including the famous theorem which bears his name.) Although I personally incline towards a Bayesian position, it is necessary to give an account of both positions. In the course of this book we shall be shamelessly eclectic. Most of the techniques we shall examine are based upon Frequentist analysis; we shall, however, consider Bayes' theorem in more detail, partly because of its intrinsic interest and partly because of the importance it has in some areas of modern psychology.

Stated in the baldest terms possible, the Frequentist view is that the probability of an event happening is a measure of the relative frequency of its occurrence in the particular circumstances specified; whilst the Bayesian view is that the probability of an event is a measure of the expectations of that event held by an ideally consistent and rational observer. Both these definitions obviously require amplification.

Suppose you threw a perfectly balanced die. Any one of six numbers can appear uppermost. Continue throwing it, again and again, each time recording the result. It will be found, as the series of throws gets larger, that the proportion of all results which are any particular face, say 6, comes ever closer to $\frac{1}{6}$ of the total. In mathematical terms, when the repetition of a process yields a quantity which approaches steadily towards some particular number, that number is called the "limit" of the process; so, in this case, we can say that in the limit, as the number of throws gets very large indeed, the proportion may be supposed to be practically indistinguishable from $\frac{1}{6}$; and it is thus stated that the "limiting" frequency of sixes is $\frac{1}{6}$. The Frequentist position is precisely that "The *probability* of a six is $\frac{1}{6}$" *means* that the frequency of sixes is $\frac{1}{6}$ of the total number of throws when that number is very large. More generally, and as we saw, the probability of an event is the limiting frequency of its occurrence.

The Bayesian will, of course, accept the argument about limiting frequencies entirely. He will, however, attach a subtly different meaning to the conclusion. If the judgment "This will certainly happen" be represented by the number 1, and the judgment "This will certainly not happen" by the number 0, he will conclude from the foregoing reasoning that the rational expectation of a six will have the measure $\frac{1}{6}$.

If we stop and think about this it will become evident that, in the case of the die, or of a coin, or indeed of a wide variety of broadly analogous cases, there will be no difference between the value given to a probability by a Frequentist and that given by a Bayesian; although they may differ somewhat in the precise meaning they attach to the term. However, it will also be evident that the Frequentist will have difficulty in attaching his interpretation to the word "probability" in some of the other uses that spring to mind. What could be the limiting frequency of John getting a first? He is only going to take the exam once. In fact it is possible (see, e.g., Nagel, 1960) to extend the Frequentist interpretation to cover a number of usages that are not obviously of its kind; but it seems very doubtful that it could be always used. A Frequentist must therefore conclude that "probability" has more than one meaning, and that measures cannot properly be extended to all of them.

The Bayesian, on the other hand, labours under no such limitation. He can, quite consistently, give a probability of $\frac{1}{6}$ for a six, and say of John "He's certainly bright enough; but hasn't been working as hard as he should; so I'd reckon the probability of his getting a first at about $\frac{1}{2}$." Suppose, further, that John's nextdoor neighbour then said "Well, actually, I think you're wrong there: John's idleness is just an act; and I happen to know that he never puts in less than six hours' hard work a day." The reply might be: "If that's so, I'd put the probability up to $\frac{4}{5}$."

We see, in fact, that *Probability is adjusted in the light of the evidence.* The optimal way of making such adjustments is a central part of Bayesian theory, which we will look at in Chapter 3.

One of the most important and direct ways of adjusting probability is statistics. At the risk of sounding a trifle circular, I must repeat that the art of statistics is derived from that branch of mathematics known as the theory of probability. This is not surprising when thought about; for, since statistics is a principal method for finding and adjusting probabilities, it must ultimately depend upon our theories of probability for its justification.

Mathematics is the most powerful instrument of human thought yet invented; but like all powerful things it needs to be handled with care and understanding. There are pitfalls in going from basic theory to real coins and dice — far more in going from basic theory to the vagaries of human behaviour. Any mathematical investigation begins by making a number of precisely stated assumptions — collectively called "the model" in contemporary jargon — and will follow out exactly the consequences of these assumptions. If, however, we attempt to apply the results of the mathematical study to a real situation which differs in some important way from the assumed conditions, we will only succeed in making fools of ourselves.

Of course, no mathematical model is *exactly* satisfied by the real world: mathematics deals with perfect circles, infinite collections, exact dimensions and suchlike idealizations which are not found in practice. Often this does not matter: no wheel is perfectly circular; but the geometry of Euclidean circles can perfectly well be used to describe or design one. Similarly, no die is perfectly balanced; but most reasonably well-made dice come close enough to this ideal for simple probability theory to cope with the results of dice games (elegantly called "crap-shooting" in the USA). And just as real rules are never geometrically straight or perfectly equal in length but quite usable, so statistical tests are roughly, and often very closely applicable.

It is important to be alert for situations which violate readily made assumptions; but some statistical tests, as we shall see in later chapters, have the fortunate characteristic of being "robust" to some violations of their assumptions. This is a term meaning that the test will still give reliable results when one of the assumptions upon which it is based does not hold. However, *no statistical technique is a substitute for thought*. Such techniques can be the most valuable aids to and tools for thought; but never are they substitutes.

A given statistical measure is referred to in the singular as "a statistic". It is the job of the person handling data to choose the appropriate statistic for his purpose, to ensure — as far as is possible — that he uses one which does not require for its validity conditions which may not be satisfied in the particular case, and to be clear about the nature and limitations of the answer he gets. It is just as important that he conveys to others precisely what his results mean, and how reliable they are. They can never, of course, be completely certain; but they can, in favourable circumstances, be good enough for a reasonable man to be prepared to risk his neck on: which is enough for all practical purposes.

I was once bullied by my son into playing a game of "Monopoly". I tended to discount his loud protests that he "never" threw a six; and patiently explained that he couldn't expect to do so more than a sixth of the time, and that, in the nature of things, he would often go for more than six throws without one. To demonstrate this, we threw a total of 120 times, tabulating the results, which were as follows: Ones, 48; twos, 20; threes, 16; fours, 17; fives, 16; sixes, 3. Now this is a perfectly *possible* result with an ideally fair die: however, it is not *likely*. Indeed, a statistic which we shall meet again in a later chapter tells me that the probability of three or fewer sixes in 120 throws is less than 1 in 1000. I cannot finally prove that the die is biased; but that probability is enough for me: I would be most unwilling to use it in any game in which I wished to see "fair play". (We don't use it any more.)

Human beings are far more "biased" than any die. A die does not remember past events, has no expectation of future ones and cannot learn anything about its surroundings. A human being, and even the lowly rat to a noticeable extent, does all these things. This we shall have to bear in mind.

References

Bayes, T. "Essay towards solving a problem in the doctrine of chances", *Phil. Trans. Roy. Soc.*, 1763, **53**, 370—408. (Reprinted: *Biometrika*, 1958, **45**, 293—315.)

Nagel, E. "The Meaning of Probability", in J. R. Newman (Ed.), *The World of Mathematics*. Allen & Unwin, London, 1960, **ii**, 1398—1414.

2

A look at the average man

One of the most basic statistical needs is simply to describe a population; and in this there are sufficient problems to have given rise to the gibe about lies, damned lies and statistics. The first question is: whereabouts is the population we are considering on the dimension we are interested in? Supposing, for example, we were interested in the intelligence of undergraduates, we might ask: what sort of I.Q. do undergraduates have? The immediate answer, of course, is that they vary a lot; and a layman would then press the query by saying: "Yes, of course, but what's the average?"

Now "average" is a tricky word. It is used to cover three technical terms, namely *mean*, *median* and *mode*, which we will define in a little while, besides a further use in the vague sense of "long term" (". . . on the average"). It is also rather vaguely supposed that these three meanings are the same; and though they sometimes are, they often are not, thereby providing ample opportunity for deception.

The time has come to introduce a little mathematical notation. Suppose a population consists of n items. Each member of this population will have a particular value in which we are interested: in the case of the undergraduates we just thought of this will be his I.Q. score. In general, this value will be a number: it might — as in this case — be points of I.Q.; in other contexts it might be time in seconds, or mass in grams, or the value of some quite complicated algebraic formula. Let the value of (or associated with) the first member of the population be written x_1 (read as "x-one", or as "x-suffix one" if there is any possibility of confusion). That of the second will be x_2, of the third x_3 and so on to the nth and last which is x_n. (We do not, at this stage, suppose that the n values have been ranked in any sort of order. The nth term is not necessarily the largest, only the last to be selected; neither is the first the smallest, only the first to be selected.) In general, any member — say the ith — is said to have the value x_i, and we can say that the population consists of all x_i where i runs from 1 to n. This is written x_i, $i = 1 \ldots n$.

The next symbol we need (tedious at this stage, but essential later on) is Σ, read "sigma" (it is the Greek capital S) which means "the sum of". The expression

$$\sum_{i=1}^{n} (x_i)$$

means "The sum of all the terms x_i, where i runs from 1 to n". If we wished, for some reason, only to sum from the 10th to the 30th term, we would write

$$\sum_{i=10}^{30} (x_i).$$

We can now write down the formula for the *mean* of the population. It is:

$$\left[\sum_{i=1}^{n} (x_i) \right] \Big/ n.$$

(This of course, is one of the common-usage "averages": you add up all the values, and divide by the number of values.)

If the instances $x_1 \ldots x_n$ constitute the whole population under consideration, the symbol for the mean is μ (Greek letter "mu" — a small "m" — read "mew", as in "cat"). If, however, $x_1 \ldots x_n$ are only a *sample* of the population (as, e.g., the heights of 1000 adult males are only a sample of the heights of all adult males) the symbol for the mean of the sample is $\bar{x}$ (read as "x bar").

Now let us suppose that a population has been ranked in order of magnitude; so that x_1 is the smallest and x_n is the largest value. (If some members are of equal value, the ordering of the equal members does not matter. If, e.g., we are ordering the height of a group of men, and we find, having ordered the first ten, that there are three whose heights are as equal as our measurements can find, we arbitrarily allot them numbers 11, 12 and 13.)

The number n (= the number of individuals in the population) will, of course, be either odd or even. In the first case, there must exist some number m such that $n = 2m + 1$; in the latter case there must exist some number r such that $n = 2r$. In the former case (n is odd) the $(m + 1)$th value in the ordered population — i.e., x_{m+1} — is called the *median*; in the second case (where n is even) the *median value* is $\frac{1}{2}(x_r + x_{r+1})$.

In most practical situations, it will be found that some values will be indistinguishable, either because measurement is insufficiently exact, or because it is convenient to use fairly coarse gradations. The value which has the greatest number of instances is then called the *mode* of the population. (This is the third common usage of "average", meaning the most "popular" value.)

It is quite often the case that the mean, the median and the mode all coincide, or are very close together. However, it is by no means invariably so, especially in interesting social contexts; and a couple of examples may help to clarify our thoughts on the subject.

The marks obtained by a class of schoolchildren in a test are shown in Table 2.1. They have been rank ordered to save the reader the trouble.

Table 2.1   Scores obtained in a test by a class of children

Child	Score	Child	Score	Child	Score	Child	Score
1	10	9	30	17	32	25	35
2	12	10	31	18	32	26	35
3	19	11	31	19	33	27	37
4	25	12	31	20	33	28	40
5	28	13	31	21	33	29	43
6	29	14	32	22	34	30	43
7	29	15	32	23	34	31	49
8	30	16	32	24	34	—	—

In this sample population n is 31, and simple addition gives $\sum_{i=1}^{31} (x_i) = 979$, whence $\bar{x} = 31.58$. In the other words, the *mean* score is 31.58, or 32 rounded to the nearest mark. Since n is odd, and clearly $= 2 \times 15 + 1$ the *median* score is that of the 16th individual; which again is 32. Simply looking at the table tells us that the *mode* is also 32.

All very simple and straightforward; and before we proceed to look at a more complex example, we shall introduce a helpful graphical device known as a *histogram*.

A histogram is a method of displaying the number, or the relative frequency, of different values, or ranges of values found in a population. Using the population of Table 2.1 above, let us mark a horizontal axis at equal intervals, each mark corresponding to a possible score from 0 to . 50. At each mark we erect a vertical line whose length is proportional to the number of scores of that value: thus, at the point 10 there will be a line 1 unit high, at 29 a line 2 units high, at 32 a line 5 units high, and so on. The resultant diagram is called a *bar histogram*; and it will enable us to see at a glance where the mode is, and roughly how the scores are distributed (of this, there will be more later). It is often more convenient, however, to take scores within certain ranges together; and in the present

Fig. 2.1   Block histogram of test scores given in Table 2.1

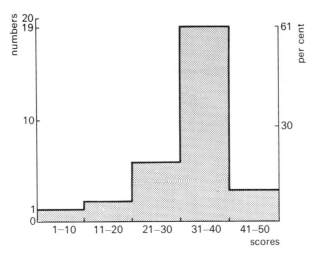

instance we might prefer to take possible scores ten at a time. To do this, we divide up the horizontal axis into equal lengths as in Fig. 2.1. The first segment corresponds to scores of 1 to 10, the second of 11 to 20, and so on. Upon each segment is erected a solid bar whose area is proportional to the number of scores falling within that range. This diagram is called a *block histogram*; and its utility will be immediately evident to the reader. The vertical scale can be either the absolute number of scores within each range, as has been marked on the left of the figure, or the percentage of all scores which fall within that range — as on the right of the figure.

Let us now proceed to consider an imaginary, but by no means utterly impossible, set of data which is not as well behaved as our first example.

Far away and long ago, the village of Gryndham Down comprised exactly 100 families. Ten families were virtually serfs; and each of these families had an income of but 10 silver pieces a year. There were forty families of day labourers, each of which managed to scrape together exactly 25 pieces a year; thirty families of peasant farmers contrived each to make 40 pieces a year; ten families of skilled tradesmen each toiled to achieve 50 pieces a year; and nine families of substantial tenant farmers each did nicely enough at 100 pieces a year. The hundredth family was that of the Wicked Baron Jasper. (If a Baron's name happens to be Jasper, he has to be wicked.) This family did not derive its income principally from extorting the hard-earned pennies of the villagers; but owned a string of brothels in various sinful cities from which an annual income of 6300 silver pieces regularly accrued.

What, asked the innocent layman, is the average income in the village? The Baron liked to put a good face on things and gave the *mean* value: the reader should do the arithmetic and confirm that this was 100 pieces a year. The day labourers, some of whom were thinking of forming a Union, quoted the modal value: the reader will at once see that this was 25 pieces a year. A visiting statistician suggested the *median*: since $n = 100$, this was $\frac{1}{2}(x_{50} + x_{51})$ and the reader will confirm that this was $32\frac{1}{2}$ pieces a year. Which of these is the best? or, at any rate, the least misleading?

It is fairly clear that the mean (which might be called the "common or journalist's average") is absurd: it suggests to the unwary a gross overestimate of the income of 90 per cent of the villagers. In the context we have created, the mode is not so bad; but it is open to the serious objection that a clear 50 per cent of the families are better off. The objection to the median is that no actual family has the income which it specifies — rather as no actual family ever has 2.2 children or whatever the "national average" may be. Nevertheless, it is probably the "best" — i.e., the least deceptive — value; since, as the reader will see, some 70 per cent of the villagers are not far off the median, leaving aside, for the moment, the exact significance of "not far off".

We have now looked at two imaginary but not at all impossible cases. In one we saw that the mean, median and mode nicely coincided; in the other they diverged so much as to be virtually incompatible as useful descriptions. It is important to understand why one involved troubles the other did not. This involves us in discussions of the distribution of a population.

The reader should sketch a bar histogram of the children's scores in Table 2.1, and of the various family incomes in Gryndham Down. (For the latter, he might find it convenient to use the logarithm of the income rather than its absolute value; if he does not the horizontal axis will be inconveniently long.) He will at once see that for the children's scores there is a close bunching around the mode, and that the distribution is not too asymmetrical about that mode. For the Gryndham Down values, on the other hand, there is very little on one side of the mode, and a very long tapering "tail" on the other: in statistical terminology, the distribution is heavily skewed.

It becomes obvious when we think about it that the mean is a reasonable answer to the "whereabouts" question when the distribution of the population is roughly symmetrical about a single mode, but highly improper when the distribution is heavily skewed. (There are methods for coping with populations with more than one peak in their histogram; but we shall not discuss these here.)

The median is usable in either case; but *must* be used in the latter.

(Unless, of course, like the Wicked Baron, you actually want to fool people. But that is a moral problem that I must leave to you.)

The reader will have noted that I have used two imprecise terms: "roughly" symmetrical and "heavily" skewed. He may well ask; how rough is "roughly" and how heavy is "heavily"? I am sorry to say that no hard-and-fast rule can be offered for answering this perfectly fair and proper question. As so often in statistics (which is a high art, not a set of rigid rules) the user must exercise his judgment. It is very rare, outside a textbook, to find populations or samples of exact symmetry and unambiguous mode. Often, however, these conditions are approached very closely indeed: the distribution of height within a human population is a good example of this. In general, a good rule of thumb is this: if the mean and the median are sufficiently different to give a divergent "first impression", use the latter. If there is any serious doubt, medians are safest.

Fig. 2.2   Histograms showing distribution of unspecified quantity in two populations

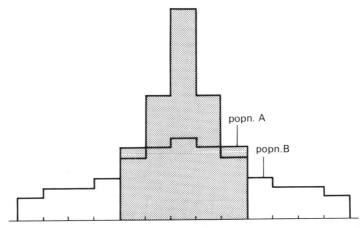

We must now proceed to consider a related problem of great importance. Look at the two histograms comprising Fig. 2.2. (Precisely what is the quality being measured need not detain us for the moment.) It is readily evident upon inspection that the two populations are identical in mean, median and mode. Are we then justified in regarding them as effectively indistinguishable? Quite clearly we are not justified in doing anything of the kind; for the two are markedly different in their distributions; and we evidently need some measure for the extent to which a distribution "spreads" about its mean or median position.

An idea which occurs fairly readily is to state the *range*. However, although we may know the range of a small available population, or of a sample, precisely, we can rarely be sure that the range of a sample includes the entire range of the parent population from which it was taken. For example, if you took a random sample consisting of 100 adult men, it is improbable that you would find in it anyone whose height was over 195 cm; but you would be wrong if you supposed this to be the upper limit of all human heights; for taller men do indeed exist. You would be on stronger ground if you stated the proportion which lay within specified bounds; and this is a common statistical practice.

The most commonly quoted ranges in practical statistical work are the *interquartile* and *90 per cent* ranges. The interquartile range specifies two bounds such that the lowest quarter of the sample is outside the lower bound, and the highest quarter outside the upper bound. In the class scores given in Table 2.1, we can approximately give these bounds as 30 and 34. Similarly, the 90 per cent range leaves 5 per cent of the sample above the upper, and 5 per cent below the lower bound.

It is sometimes inconvenient to use either of these traditional ranges; and any other reasonable value may be chosen to suit a particular case. The reader might agree that in discussing Gryndham Down we could reasonably state that 70 per cent of the families have incomes within the range ±13 pieces (± is read "plus or minus") of the median.

A more refined measure is usually used in conjunction with means. This is known as the "standard deviation": it can be used so long as the population has a single mode and is fairly symmetrical; although, as we shall see, care has to be taken in making use of some of its properties.

Consider again a sample x_i, $i = 1 \ldots n$, with a mean $\bar{x}$. In general, some particular value x_i will differ from the mean, this difference being $(x_i - \bar{x})$, called the deviation from the mean. The more "spread" the sample or population has, the larger these values will tend to be. Some, however, will be negative, so it is convenient to use the *square* of the deviation $(x_i - \bar{x})^2$. If we consider all n deviations, it is clear that the mean square deviation, $\dfrac{1}{n} \sum\limits_{i=1}^{n} (x_i - \bar{x})^2$, will be larger the more spread there is in the population, and smaller the less spread there is. The mean square deviation is usually called the *variance* of the population. We have been discussing thus far a wholly known population, or a sample. If we are using a sample to estimate the variance of a population, it is more usual to use

$$\text{variance} = \frac{1}{(n-1)} \sum_{i=1}^{n} (x_i - \bar{x})^2 \tag{2.1}$$

because (as we saw in the case of the example of the sample of heights) a sample variance is likely to underestimate the variability somewhat. The variance is itself sometimes used as a measure of the spread. However, it is more usual to use the square root of the variance

$$\left[\frac{1}{(n-1)} \sum_{i=1}^{n} (x_i - \bar{x})^2 \right]^{\frac{1}{2}}$$

which is known as the *standard deviation.*

(There are a number of reasons for doing this, the most straightforward being that it keeps all the units in line. For example, if x_i be measures of length in centimetres, then $(x_i - \bar{x})$ is also in centimetres; but $(x_i - \bar{x})^2$ — and hence the variance — is in square centimetres: units of area. By taking the standard deviation, however, the measure of spread is now consistent with the mean and the individual measures, in centimetres.)

If the population is wholly known, the standard deviation is generally denoted by σ, the lower-case Greek letter sigma, or small s. In the more usual case of samples, however, the standard deviation is usually denoted by the ordinary s, and the variance by s^2.

It is often convenient to use a slightly different form of the expression for s^2, for the sake of ease of computation. This is:

$$s^2 = \frac{n}{n-1} \left[\frac{1}{n} \sum_{i=1}^{n} (x_i^2) - \bar{x}^2 \right] \tag{2.2}$$

The proof that this is equivalent to 2.1 is simple, and we will now state it. If, however, the reader prefers to take it on trust (and I will not fool you) he may skip and rejoin us at the paragraph marked ** below.

$$\frac{\Sigma (x_i - \bar{x})^2}{n} = \frac{\Sigma (x_i^2)}{n} - \frac{2\bar{x} \Sigma (x_i)}{n} + \frac{\Sigma (\bar{x}^2)}{n}$$

but $\quad \dfrac{\Sigma (x_i)}{n} = \bar{x}$ (by definition); and $\Sigma (\bar{x}^2) = n(\bar{x}^2)$

therefore

$$\frac{\Sigma (x_i - \bar{x})^2}{n} = \frac{\Sigma (x_i^2)}{n} - 2\bar{x}^2 + \bar{x}^2 = \frac{\Sigma (x_i^2)}{n} - \bar{x}^2$$

and the term $\dfrac{n}{n-1}$ is needed to adjust from σ^2 to s^2.

**If we refer back to the data of Table 2.1, we can see quite straight-forwardly how to use this formula to compute s. We tabulate the data in columns (only the first and last few entries are shown). Thus:

i	x_i	x_i^2
1	10	100
2	12	144
----	----	-----
29	43	1849
30	43	1849
31	49	2401
Σ:	979	32,613

Clearly the total of the x_i column is Σx_i, that of the $(x_i)^2$ column is $\Sigma(x_i^2)$. Dividing the first by n (31 in this case) gives $\bar{x}$; dividing the second by n gives $\dfrac{\Sigma(x_i^2)}{n}$. (If the reader happens to have access to one of those handy desk calculators which sum numbers and squares on separate registers, this is child's play.) Carrying out the division, and rounding to the nearest unit (the reader should check the sums himself) gives:

$$s = \sqrt{(1052 - 32^2)} = \sqrt{(1052 - 1024)} = \sqrt{28} \doteq 5.3$$

The standard deviation, or s.d. for short, acquires its greatest importance in connection with a particular form of population distribution, sometimes called the Gaussian distribution (after the great mathematician K. F. Gauss who studied it) and sometimes misleadingly called the Normal Distribution for historical reasons we shall shortly explain. We must now proceed to consider this interesting distribution, and first we must slightly revise our nomenclature.

So far we have used x, variously suffixed to denote values of whatever variable — be it height, marks, or whatever — we are concerned with that occur in the population we are studying. For the rest of this chapter, any *observed* value will be denoted by a straight x (like that), while the ordinary x will refer to any possible value of the variable, observed or not.

Now we must consider a refinement of the block histogram we have already looked at. If our measurements were capable of indefinitely fine distinctions, and our population very large, we could make the bounds of our vertical blocks as close together as we pleased (see Fig. 2.3a). In the limit — using this phrase in the sense employed in Chapter 1 — the irregular staircase formed by the tops of the blocks will be indistinguishable from a smooth curve. The final form of a population histogram would

Fig. 2.3 (a) The refined histogram tends to (b) the probability distribution

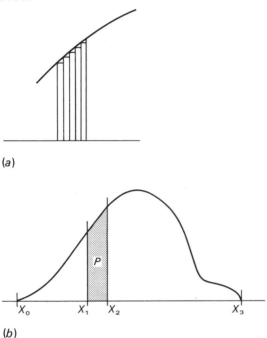

(a)

(b)

look something like the curve of Fig. 2.3b, but in the process it can be made to undergo a subtle change in significance which we must again be clear about.

In the original block histogram, the width of each block was pre-set; and the area, and hence the height, was made proportional to the number of x_i falling within that range. Thus the area of each bar was proportional to that number, and the total area of the bars was proportional to n — the total number of individuals in the population. Similarly the ratio of the area of any bar to the total area was proportional to the fraction of the total population lying in that range.

If, now, in Fig. 2.3b we *define* the area bounded by the x-axis and the curve as 1 unit, then since, in the figure we have drawn, all values in the population lie between x_0 and x_3, the probability of any member of the population having a value in this range is, by mere tautology, also 1. Behold! We have turned our curve into a *probability density function*, to give it its technical name. If we take some area, such as that shaded in the figure, which is bounded by the curve, the axis, and two verticals erected on the axis at x_1 and x_2, then the area thus formed, say P, will equal the fraction of the total population lying between x_1 and x_2. Hence the shaded area P will be equal to the probability that any randomly chosen x_i is between x_1 and x_2.

[Strictly for the benefit of those familiar with calculus notation, if the density curve is $y = f(x)$, then for any fixed values A, B, the probability that any randomly chosen x_i is such that

$A \leqslant x_i < B$ is

$$P = \int_A^B f(x)\,dx \quad]$$

The ground now being cleared, we can approach the Gaussian curve.

Early in the eighteenth century, astronomers became aware that, if a large number of observations of some fixed quantity were made — say of the angular separation of two fixed stars — the various readings were not all identical. This was natural enough; for numerous sources of error were present: the observer might make a small error in reading his instrument, he might be a trifle early or late in deciding just when a particular star coincided with his cross-wire, there would certainly be variable atmospheric disturbance, and so on. It was found that, if the probability density curve of all the readings of some value were plotted, it approximated at last the form in Fig. 2.4: there was a marked preponderance of readings around the mean value, with a symmetrical falling away on both sides giving a characteristic bell-shape. This curve became known as the "Normal Error Curve", for the odd reason that it was in fact the normal distribution of errors.

Any reader likely to be unnerved by calculus notation may again skip the next passage, and, without very much loss, rejoin us at the point marked **.

Fig. 2.4   The Gaussian or normal distribution

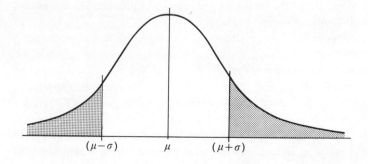

It follows from the general form of the curve that, if it be described as $f(x)$ (plotted against x) then:

$$\frac{df(x)}{dx} \begin{cases} = 0 \text{ at } x = \bar{x} \\ > 0 \text{ for } x < \bar{x} \\ < 0 \text{ for } x > \bar{x} \end{cases}$$

and that $f(x) \rightarrow 0$ as $x \rightarrow \pm \infty$.

It was early realized that these general conditions were satisfied by an expression of the form

$$y = A \exp[-K(x - \bar{x})^2]$$

when A and K are constants.

**K. F. Gauss showed that, if large numbers of small, independent terms were contributing to the errors, the curve would be expected to be of a form described by the expression (do not be frightened by its fierce face):

$$y = \frac{1}{\sigma\sqrt{(2\pi)}} \exp\left[-\frac{1}{2\sigma}(x - \mu)^2\right]$$

and that any particular example could be precisely defined by the terms μ and σ, the population mean and s.d.

It gradually became evident that this function has surprisingly wide applications — though not, I fear, as wide as some would seem to think — and a number of very interesting and useful properties.

To mention a few applications only: the distribution of human heights is very nearly Gaussian, so is I.Q., so are human sensory abilities. So likewise (a point we will examine in more detail later) are numerous statistical sampling measures. So are *not* things like income (even outside Gryndham Down) or human response times. Once again, you see, a very useful tool has to be used with discretion. Let us now look at some of the useful properties of this tool.

The most important property is this: once we know the mean and the s.d., we know what percentage of the total population lies between *any* specified limits. For example, in Fig. 2.4, the unshaded area lying between $(\mu - \sigma)$ and $(\mu + \sigma)$ contains 68.3 per cent of the total area, and hence 68.3 per cent of the total population *whatever μ and σ may happen to be*. Further, since the distribution is symmetrical, we know that each of the two shaded "tails" contains $\frac{1}{2}(100 - 68.3)$ per cent of the population — i.e., approximately 15.9 per cent. Since the die-away is rapid, only 2.3 per cent lies in either tail more than 2 s.d.s distant from the mean.

These factors enable us to give what are called "confidence limits" for any Gaussian population whose parameters (μ and σ) are known. For example, if a machine is making bars of chocolate whose mean mass is 175 g, with an s.d. of 2 g, we can be 95 per cent sure — i.e., we can properly assign a probability a shade higher than 0.95 — that any given bar weighs more than 171 g and less than 179 g.

Theoretically, the curve extends infinitely in either direction; in practice, of course, this does not happen: as noted, human height is almost exactly Gaussian in its distribution; but neither Brobdingnagians nor Lilliputians are, in fact, possible: and no man can ever have any negative height. In practice, therefore, we neglect sufficiently small probabilities: what "sufficiently small" is, depends on particular circumstances.

It is possible to take advantage of the symmetry of the Gaussian distribution to construct tables, such as Table A in the appendix here, which enable us to find the proportion of the area lying between any two values. We know that 50 per cent of the area lies to the left of the mean. If, then, we measure all distances from the mean as multiples of the s.d., we can tabulate the total area to the left of any given point. Thus, e.g., the table tells us that 95.5 per cent of the total area lies to the left of a point 1.70 s.d.s to the right of the mean; hence the "tail" beyond this point contains 4.5 per cent of the total area. Also, by symmetry, 4.5 per cent of the total area lies in the corresponding tail 1.70 s.d.s to the left of the mean, and hence 91 per cent lies within ±1.70 σ of the mean; and this is true whatever values μ and σ may have.

The usefulness of this kind of tabulation can become clear if we consider a couple of examples. First a purely numerical one: let a Gaussian distribution have a mean of 50 and an s.d. of 10. What is the probability of a randomly chosen individual from the population lying between 40 and 65?

Forty is 1 s.d. from the mean; and from the table, 84.1 per cent of the total area lies between this value and the opposite tail indefinitely extended. Hence 100 − 84.1 = 15.9 per cent of the total lies in the tail consisting of all $x_i < 40$. Similarly, 65 is 1.5 s.d.s from the mean, and the other tail cut off including all $x_i > 65$ contains 6.7 per cent of the total. The two tails cut off by the chosen bounds therefore comprise together (15.9 + 6.7 per cent) = 22.6 per cent of the total area. Therefore 77.4 per cent of the total area lies between 40 and 65; and hence the desired probability is 0.774.

A second problem: suppose you are a manufacturer of colanders to be used for draining peas. You might decide that your customers would realise that some peas are very small and are bound to fall through the holes, but would be satisfied if not more than 1 per cent of these delectable vegetables were lost in this way. If your research department told you that the distribution of diameters of peas is Gaussian with a mean of

μ cm and an s.d. of σ cm, you can easily meet this demand. Appendix Table A tells us that 99 per cent of a population lies between one in-definitely prolonged tail and a point 2.33 σ the other side of the mean. You must therefore ensure that none of your machines stamps holes with diameters greater than (μ − 2.33 σ) cm.

In this chapter we have had a look at the various usages of "average", and have tried to see how these are best used. The popular unstated premise that mean, median and mode coincide, has been seen to be erroneous; and we have further seen how utterly confusing this fallacy can be. We have seen that median values are most useful when the population is skewed; and that, whether mean or median is used, some indication of range is essential if we are to get a just estimate of the state of affairs represented by our data. When a population is roughly Gaussian this is very convenient, as it enables us to estimate the proportions lying within specified bounds very easily.

An average Englishman is doubtless 173 cm tall, has 2.2 children, an I.Q. of 100 and an income of £x/year. (I have been too lazy to look up that x; and in any case it will have changed by the time you read this.) All these are *means*: right enough for the first and third, more than doubtful for the second, downright deceptive for the fourth. But we now know what to look out for when we see statements of this kind.

This chapter, and each succeeding chapter, ends with a few exercises for the reader, who is earnestly recommended to have a stab at them. Afterwards (but please not before) he can turn to page 145 where he will find outline answers and comments.

Exercises

1. In Good King Charles' Golden Days the mean age at marriage was older than the mean age at death. Does this mean that most persons were dead when they got married? If not, explain what it does mean.
2. Anthropometric data give the mean height of adult males as 173 cm with an s.d. of 6.5 cm. If you wish to ensure that 99 per cent of adult males should be able to walk through doorways without having to duck, how high should the frames be made?
3. Given that I.Q. is pretty well Gaussian in its distribution ($\bar{x} = 100$; $s = 15$) and that Universities accept only persons in the top 8 per cent of the population, what (referring back to the hypothetical question at the beginning of the chapter) would be the most suitable measure for the "average" student I.Q.? And what do you (roughly) estimate it to be?
4. Members of a particular human group X have been studied for some genetically determined quality Q. It is found that the distribution of Q-scores is roughly Gaussian with a mean of 216 and s.d. 15. Another

group Y has been similarly studied; and their corresponding values are found to be 196 and 15. Interesting and readily observable characteristics C are found with any individual whose Q-score is below 180. What can you infer from these data? Comment.

3

Some basic notions on probability

Before we go on to some more aspects of practical statistics, it is necessary to clarify a few basic ideas on probability. As before, we shall require some mathematical notation on the way; but this is for simplicity's sake. People often refer to mathematical statements as "complicated"; but this is a mistake. Unfamiliar, and hence puzzling, they may be; but they are useful precisely because they are such concise and unambiguous methods of statement: translating a mathematical statement into ordinary words invariably leaves us with something much longer than the original, and far less precise (try it yourself with something really simple, like $y = 2x^2 + x - 5$).

Let us, then, start by declaring that A shall represent some event. The probability of that event we shall write $p(A)$. As we have seen, $0 \leqslant p \leqslant 1$. If $p = 0$, it may be taken that A is certain not to happen; if $p = 1$, A is certain to happen: most interesting events lie between these extreme cases.

In Chapter 1 we saw that the value of a particular $p(A)$, namely the probability of John getting a first, depended very strongly upon the nature of our information concerning John — naturally enough. Let our relevant information be H; then we should really say — in any practical situation — not simply "the probability of A", but "the probability of A given that H is true". This we write $p(A|H)$ (read "pA given H"), and it is called the *conditional probability of A given H*.

In all practical situations, of course, probabilities are conditional probabilities. We do not always preface our estimates with some such phrase as "in the light of what we know about . . ."; but it is invariably implied. The reader should bear this in mind, because we shall only use the formal symbolism when considering evidence leading to change in H; for the most part we shall simply write $p(A)$.

Three laws of probability

Suppose that A and B are any two mutually exclusive events. (For example, in throwing a die, A could be the outcome "6", and B the outcome "5"). The probability of the combined event "either A or B" is the sum of the several probabilities. This statement is sometimes called the *First Law of Probability*. The symbolism $A \cup B$ is often used for "A or B"; and the law is then written $p(A \cup B) = p(A) + p(B)$.

The law is readily extended to any number of mutually exclusive events. Thus:

$$p(A \cup B \cup C \ldots \cup N) = p(A) + p(B) + p(C) + \cdots + p(N) \ldots \quad (3.1)$$

The proof of this is very simple. Since each of the two events in the original statement can be *any* two mutually exclusive events, we can, when considering the case of three such events A, B and C, take two of them together and name the outcome $(A \cup B)$ as outcome X. Then the law states that

$$p(X \cup C) = p(X) + p(C)$$

but already

$$p(X) = p(A \cup B) = p(A) + p(B)$$

therefore

$$p(A \cup B \cup C) = p(A) + p(B) + p(C).$$

In the same way, we could now name the combined outcome $(A \cup B \cup C)$ as outcome Y; and the result follows for four events; and so on for any number. (This kind of proof is known as proof by recurrence relations.)

A special case of this law, and one which is by no means trivial, is when one of the two events is the negation of the other — i.e., the event which must happen if the other does not. Clearly the die example I just offered is not such a case; for the outcome can be other than 5 or 6. However, suppose I take event A to be that, this time next year (I am writing in July 1972), Richard M. Nixon will be President of the United States, and B to be that he will not. Clearly one or the other must be true: there is no third state available. (The reader will know which it is by the time he sees this.) Thus $p(A) + p(B) = 1$. Here we may introduce another handy sign, and, in this special case, write $\bar{A}$ (read "not A") instead of B. Thus:

$$p(A) = 1 - p(\bar{A})$$

or equivalently:

$$p(\bar{A}) = 1 - p(A)$$

It is very important to remember that the law stated in (3.1) and its corollary just given apply to *mutually exclusive* events. If I put my hand into a box of chessmen and take out the first piece my fingers touch, I can assert that the probability of its being a pawn is $\frac{1}{2}$, and that the probability of its being white is $\frac{1}{2}$. But, clearly, the probability of its being white *or* a pawn is not 1! These outcomes are not mutually exclusive, since half the white pieces themselves are pawns.

To say that two events are not incompatible is equivalent to saying that both may occur. The statement which covers this eventuality is this: if A and B are two events which are not mutually exclusive, then the probability that both will occur is equal to the product of the probability of A and the probability of B given that A has occurred. This is the *Second Law of Probability*, and again another sign and some clarification are called for.

The outcome "Both A and B" is often written $A \cap B$ (read "A and B"), and the second law can be written:

$$p(A \cap B) = p(A) \cdot p(B \mid A) \ldots \tag{3.2}$$

Note that, in this case, we have written $p(B \mid A)$, not merely $p(B)$. This is necessary, because events need not be independent; and the fact of event A having happened can affect the probability of another event B. Often, of course, this is not so, as in the case of tossing coins. If B is independent of A, then $p(B \mid A) = p(B \mid \overline{A})$. The "gambler's fallacy" turns upon the non-appreciation of independence. (This fallacy, also known as the "fallacy of continued probability", is the extraordinarily widespread belief that a run of heads makes a tail more likely next time. In fact the outcome of a particular toss of a coin is unaffected by the result of a previous toss.)

In many practical situations, however, events which interest us are not truly independent, or even nearly so. At the time I am writing, politically interested persons in Britain are asking such questions as "What are the chances that this government will succeed in its economic policy (A) and also win the next election (B)?" Clearly, in this case $p(B \mid A)$ is very much greater than $p(B \mid \overline{A})$.

Just as with the first law, and using the same mathematical device, the second law can be extended to any number of events. In this case:

$$p(A \cap B \cap C \ldots M \cap N) = p(A)p(B \mid A) \cdot p(C \mid A \cap B) \ldots p(N \mid A \cap B \cap C \ldots \cap M)$$

The reader might like to essay the proof for himself.

A very important conclusion which may be drawn from the second law is the enormous superiority, in terms of reliability, of systems in which any one of a number of sub-systems *can* suffice to systems in which all of a number of sub-systems *must* function. Imagine a top-security establishment, which is guarded successively by a photo-electric burglar alarm, a guard dog, an infra-red sensor and a concealed trip-wire. The photocell, infra-red equipment, and trip-wire are quite separate, and the dog knows nothing of any of them. Let us further suppose that the chance of an ill-intentioned person evading the photocell is 0.1, the infra-red equipment again 0.1, the trip-wire 0.2 and the dog 0.25. The probability of his affecting a successful entry is, of course, equal to p(all systems fail), which, by the second law, is $0.1 \times 0.1 \times 0.2 \times 0.25 = 0.0005$. However, suppose that, instead of the four systems functioning independently, there was an overall alarm which was only triggered when all four systems together recorded the presence of an intruder. He would then be apprehended only if all four systems succeeded, and the probability of his entry would be $1 - p$(four successes) = $1 - (0.9 \times 0.9 \times 0.8 \times 0.75) = 1 - 0.486 = 0.514$.

That is to say, he would be a *thousand times* more likely to get in; indeed the balance of the odds would be slightly in his favour. The moral of this is obvious.

The reader will recall that, in Chapter 1, we discussed the hypothetical question of whether or not John would get a first. Suppose that, during the discussion of John's prospects, someone had asked, "What is the chance of his getting that job with X?", the answer might very well be "That really depends on whether or not he gets that first." This indeed is typical of many real situations: the uncertainty of some event (in this case, John getting the job) is itself dependent upon the outcome of another uncertain event (in this case, the class of John's degree), for which the possibilities are both mutually exclusive and exhaustive. Let A and $\overline{A}$ be these two antecedent possibilities, and B the dependent uncertain event that interests us. Then:

$$p(B) = p(B \mid A) \cdot p(A) + p(B \mid \overline{A}) \cdot p(\overline{A}) \qquad (3.3)$$

This is the *Third Law of Probability*. (Note, by the way, that we have already reached a statement which it would be absurdly tedious to express in words.)

Following on the discussion of John's career, we might make some statement like this: "John is a good candidate; but a lot of good people will be trying for that job. If he gets that first, I'd put his chance at $\frac{1}{3}$; but if he doesn't I wouldn't put it better than $\frac{1}{10}$." Now we had already decided that John's chance of a first was $\frac{4}{5}$. (He is, you will recall, a very bright lad.) So the third law gives the probability of his obtaining this job as $p(\text{Job}) = \frac{1}{3} \cdot \frac{4}{5} + \frac{1}{10} \cdot \frac{1}{5}$ which is $\frac{43}{150}$ or rather over $\frac{1}{4}$: not very good, but not negligible either.

Bayes' Theorem

Let us now pursue the saga of John's fortunes a little further. A few weeks have passed; and we happen to meet a friend of John's, who says, in the course of conversation: "I was glad that John got that job with X." We are glad to learn this (whatever de Rochefoucauld says) and go on our way rejoicing. However, we were so pleased that we forgot to ask about his degree results; and we are therefore still uncertain whether or not he got that first. By now, of course, it is not so important; but, on thinking it over, we decided that we must revise our estimate of the likelihood of that coveted first.

Our prior probability for the hypothesis "John will get a first" was, as we saw, $\frac{4}{5}$. However, we also agreed that the probability of "first *and* job" was very much greater than that of "no first *and* job". We now have an extra datum, namely, that he did get the job; so how should we revise our estimate of the probability that the class of John's degree was a first?

The fate of John is not a matter of vast importance to more than a few persons; but the general form of the question we have raised is so important that we should sit back and think about it before proceeding to its solution.

The question is this: given that a certain prior probability has been attached to a statement, how should that probability be adjusted in the light of some new evidence? This is a matter of the last consequence. One of the basic problems, not merely of scientists in their professional capacity, but of all rational men, is *to adjust opinion in the light of evidence*.

Before proceeding to the formal investigation of this question, let us think for a moment, in quite general terms, of the *sort* of answer that would be reasonable. If we make an observation that is consistent with our opinion, but which is quite likely to happen whether our opinion happens to be true or false, our confidence in that opinion will not be greatly increased. On the other hand, if our present view leads us to expect some event which, on any other view, would be extremely unlikely, it would greatly strengthen our belief if we then observed this event. In other words, the confirming power of an observation is less when our prior estimate of its general probability is greater: confirming power is *inversely* related to prior probability.

It is similarly the case that we would be likely to change our opinion if we observed some event which that opinion held to be very unlikely. In the strongest case, we would have to drop our opinion if we observed something which it absolutely ruled out. The *disconfirming* power of an observation, then, is greater the *less* its likelihood according to the opinion we are testing.

Naturally, also, we are more easily convinced about ideas which there are already strong grounds for believing. This is quite reasonable; and we may re-state this as saying that more probable opinions are more readily confirmed.

Thus we have, when we think about it, a general idea of how we can rationally modify our opinions: can we formulate a sharper statement of the process? We are now in a position to tackle this vital question, and to provide a solution for all cases where numbers can be attached to the probabilities. Though the proviso is by no means a negligible one (we shall discuss it further in a moment) this is a very important step indeed.

Let H and D be two events, then (3.2) gives:

$$p(D \cap H) = p(H) \cdot p(D \mid H)$$

therefore

$$p(D \mid H) = \frac{p(D \cap H)}{p(H)} \tag{3.4}$$

(We must, of course, assume that $p(H) \neq 0$.)

Now whatever D and H are, it is obvious that "D and H" is the same as "H and D", so evidently

$$p(D \cap H) = p(H \cap D) = p(D) \cdot p(H \mid D)$$

Substituting this in (3.4) gives:

$$p(D \mid H) = \frac{p(D) \cdot p(H \mid D)}{p(H)}$$

whence:

$$p(H \mid D) = \frac{p(H) \cdot p(D \mid H)}{p(D)} \tag{3.5}$$

This is *Bayes' Theorem*, one of the most important statements in the whole of probability theory and, indeed, some would claim, in the whole philosophy of science. If we let H stand for the Hypothesis, and D for the "New Datum", then $p(H \mid D)$ is the probability of the hypothesis in the light of that new datum; and the right-hand side of this splendid expression tells us how to compute it. $p(D)$ is understood to be the *prior* probability of D — i.e., the estimate we made of its probability *before* actually observing it.

The reader will see that (3.5) exactly conforms to the general ideas we discussed before setting out on the formal derivation. $p(H \mid D)$ does indeed vary inversely with $p(D)$: the less likely D is, the greater is its confirming power. $p(H \mid D)$ does decline the smaller $p(D \mid H)$ is, and falls to zero when this is zero; we feel less confidence in H the less likely it made D to appear beforehand. Also, we allow our prior confidence in H weight in the final estimate.

Back to John. Our H was the supposition "John gets a first", and we saw that $p(H) = \frac{4}{5}$. Our D was the datum "John gets the job", and for this our prior value (as we said, the probability before we knew) was $\frac{43}{150}$. We had also assigned to $p(D|H)$ — i.e., the probability of the datum given the hypothesis, in this case the probability of John getting the job if indeed he got a first — the value $\frac{1}{3}$. Thus Bayes' Theorem now adjusts our probability to:

$$p(H|D) = \frac{\frac{4}{5} \times \frac{1}{3}}{\frac{43}{150}} = \frac{40}{43} = 0.93.$$

We are now fairly confident that John did get his first (good for John). However, we are not *certain*.

This last is a very important point, both philosophically and in practice. If H is a *general* hypothesis, $p(H|D)$ can never equal 1: for even though our hypothesis clearly predicts D, we can never wholly rule out the possibility that some other factor which we haven't thought of, or don't or can't know about, may also produce the finding. However, $p(D|H)$ *can* equal zero; and hence we can be left with the certainty that our hypothesis was false. Of course, confirmation can be very strong indeed; it can be shown (though the proof would be out of place here) that repeated applications of Bayes' Theorem can eventually lead us to assign to an hypothesis a probability $p = 1 - \epsilon$ where ϵ is as small as we please. For example, we are all pretty sure that the world is round. We cannot finally and utterly prove that some fantastic system of optical or psychological illusions is not systematically deceiving us; but in practice we neglect this particular ϵ with complete cheerfulness and rationality.

It is easy to imagine disconfirming instances. You will recall that Archimedes concluded that the sun and stars moved around the earth as centre, on the perfectly rational ground that no stellar parallax (the slight apparent shift in direction of stars seen from the earth as the earth moves round the sun) could be observed — as, indeed, it could not be with the equipment then possible. However, in a strange incident described in a lost manuscript, Archimedes met a traveller from Tau Ceti III — an ancestor of the friend we met in Chapter 1. The traveller loaned him really advanced optical equipment; and noted that he had formed some quite clear prior probabilities concerning the hypothesis (H) that the earth was at the centre of the universe, and that parallax (D) might be observed. According to Archimedes $p(H)$ was about 0.9, and $p(D)$ about 0.1; $p(D|H)$, of course was simply 0: if everything rotates around us, there can be no parallax. He then made the observations, using the borrowed equipment, and consequently dropped the hypothesis: it had been falsified. Unfortunately, as I said, the manuscript enshrining this historic meeting has been lost.

In this discussion we have been treating H as a general hypothesis. In particular cases, of course, it is possible to make inferences which are certain. If we are *given* that one of two boxes contains only black balls, and the other only white ones, extracting a single ball from either will render the situation fully known. In a less artificial situation, if a man's head is cut off, his death may be taken as certain.

At this stage, it may be as well to look at some problems, so as to clarify our ideas.

To start with, here is a classic problem due to Bertrand. There are three boxes: one contains two black balls, one contains one white and one black, the third contains two white. You extract a ball at random from one of the boxes chosen at random, and find that it is black; what is the chance that the second ball in that box is black also?

Let us take (H) to be "The second ball is black", and D to be the observation that the first ball drawn was. Clearly $p(D|H) = 1$, and the prior value of H is the probability that the chosen box is the one with two black balls $= p(H) = \frac{1}{3}$. What prior value should be assigned to $p(D)$? The third law (3.3 above) clearly gives $p(D)$:

$$p(D) = \tfrac{1}{3} . 1 + \tfrac{1}{3} . \tfrac{1}{2} + \tfrac{1}{3} . 0 = \tfrac{1}{2}$$

Then Bayes' Theorem gives

$$p(H|D) = \frac{\tfrac{1}{3} . 1}{\tfrac{1}{2}} = \tfrac{2}{3}$$

There is a certain artificiality about problems of coloured balls in boxes; though much can be learned from them. Let us consider an imaginary but more realistic case.

A brilliant doctor has invented a device for screening persons suspected of suffering from the Festering Gut-Rot, a dread disease from which one person in every 1000 suffers. However, the device, though very good, is not infallible: if an individual is a sufferer, the probability of the device giving a positive response is 0.9; if he is not, there is still a probability of 0.01 that the device will respond positively. If the device records a positive response when a particular individual is screened; what is the probability that he is indeed a sufferer?

Here, let H be the hypothesis that the individual has the Festering Gut-Rot, and clearly the prior is $p(H) = 0.001$; the datum (D) is a positive response; and we have seen that $p(D|H) = 0.90$. The third law can again give $p(D)$:

$$p(D) = (0.9 \times 0.001 + 0.01 \times 0.999) = 0.011.$$

Thus

$$p(H|D) = \frac{0.9 \times 0.001}{0.011} = 0.08.$$

This result is an interesting one. Firstly, it is one of the very small class of probability problems in which laymen tend to *over*-estimate probability — as, indeed, I expect the reader did at first sight. As a general rule, those who are not compulsive gamblers tend to under-estimate probability grossly. This, however, is a psychological question rather outside our scope: the reader who is interested will find the data set out in Slovic and Lichtenstein's review paper (1971). More relevant from our point of view is the question of how best to improve the test: should the designers concentrate on reducing the already low (10 per cent) omission rate, or on reducing the already low (1 per cent) false positive rate? (an "omission" is a failure to give a correct positive response; a "false positive" is a positive response given in error).

As a general rule†, it is hard to improve one without spoiling the other; so let us examine a couple of possible changes. Suppose that the screening device were improved to a point where the probability of detecting a real sufferer were raised to 0.99, but at the cost of increasing the false positive rate to 0.02. What now is the probability that an individual who gets a positive response is indeed suffering from FGR? The reader should do the sums himself, and (if mine are right) he should find that, in this case, $p(H|D) = 0.047$. Suppose on the other hand, that it were possible to reduce the false positive rate to 0.001, but at the cost of reducing the probability of a genuine detection to 0.8; what would the probability be then? The reader will confirm, I hope, that $P(H|D) = 0.44$ — i.e., more than five times the initial value, and almost ten times the other alternative change.

Now are we to conclude that in cases of this kind it is better to reduce false positives than to reduce omissions? By no means; for here again we come upon questions which are outside the scope of statistics proper, and which involve the consequences of actions, or, in the modern jargon, the "payoff". We might well decide that the pain and wretchedness which FGR causes the sufferer, and the trouble to which a case puts many others, make it well worth while to re-examine, quite fruitlessly, nine or ten healthy persons for every real victim. On the other hand, we might, sorrowfully, decide that our available medical resources simply did not allow this; and that we must be content with the lower rate of successful prevention.

† Murphy's Law, also attributed to Sod.

The reader might imagine — indeed, it is, I think, the most widely held view — that such questions are quite outside the scope of analysis in quantitative terms. This, however, is not the case: each of us can, and in fact does, weigh this kind of "imponderable" in the most simple numerical and even monetary terms. Which of us has not decided, at one time or another, that "I'll go to that concert if I can get a ticket for £1.50; but I'm damned if I'll pay £5.00"? Is this not measuring the pleasure of music in cash terms? In principle this argument can be extended over a very wide area; and affords, perhaps, one of the most rational methods of seeking agreement on such disputed questions.

Fascinating though this question is, however, I must again urge the reader to pursue it for himself (see, e.g., Lindley, 1971): our concerns lie in narrower fields.

In this narrower field, we have a clear lesson to be learned from our screening fable. It is this: if a test is made for a *rare* event, the false positive rate can be more important than the omission rate in determining the power of the test. (The "power" of a test is, roughly, its ability to reduce our uncertainty. See Chapter 4.)

Permutations and combinations

Before we leave these basic questions, there is another matter which has to be dealt with. Often, in applied statistics we are concerned with the *ordering* of events and magnitudes. Is this sequence an improbable one? Here are a number of independent sets of rankings: just how many different rank orderings *could* there be? and so on. We need to know how many different ways there are of ordering a given set of events or objects, and how many different ways there are of selecting a given number of events or objects from some larger set.

Suppose four persons are to sit in four chairs set around a table. If they sit where they please, one at a time, the first sitter has a choice of four places; the second then has three places to choose from; the third has two; and the last man has no choice, only one chair being still unoccupied. It is evident that any second choice can be combined with any first choice; so the number of different ways of placing the first two persons is 4 × 3. By the same token, the number of different ways of placing all four persons is 4 × 3 × 2 × 1. Generalizing this argument, the number of possible arrangements of *n* objects is:

$$n(n-1) \cdot (n-2) \ldots 3.2.1$$

This quantity is called "factorial *n*", and is written *n*!. (If the reader wishes to demonstrate his familiarity with mathematicians' private jargon, he may read this as "n-shriek".)

The problem we have just solved is that of the PERMUTATION of n things, n at a time. Supposing that, at the table we started with, there were only three chairs for four persons, so that somebody had to be left standing. Evidently, the total number of arrangements of sitters would be $4 \times 3 \times 2$; and this is the number of permutations, that is of different arrangements, of four things taken three at a time. The notation for this is $^4P_3$. The reader will be able to satisfy himself that ^nP_r, where n and r are integers such that $n > r$, is

$$^nP_r = n(n-1) \cdot (n-2) \ldots (n-r+1)$$

Or more simply,

$$^nP_r = \frac{n!}{(n-r)!} \tag{3.6}$$

In a question of *arrangements*, AB is clearly different from BA. However, we are not always interested in the *ordering* of objects. Suppose we had four items A, B, C and D, and wished to know how many different pairs of items, *irrespective of order*, could be selected from them. In this case AB is the *same* selection as BA; and the number of different selections is six: AB, AC, AD, BC, BD and CD. Each such selection is called a COMBINATION. The number of combinations of four things taken two at a time is written $^4C_2$.

Now it is evident that any *combination* can give rise to a number of *permutations*. From the combination AB we can make the permutations AB and BA. Indeed, as we have seen, r things can give $r!$ permutations. This enables us to find the general value of ^nC_r. Since each combination of r things can give $r!$ permutations, it follows that:

$$^nC_r \cdot r! = {}^nP_r$$

therefore $$^nC_r = \frac{^nP_r}{r!}$$

that is $$^nC_r = \frac{n!}{(n-r)! \, r!} \tag{3.7}$$

Some words of caution must be inserted here, because common usage and mathematical terminology are in flat contradiction at this point. What locksmiths call a "combination" lock is what we must call a "permutation" lock; so far from being disregarded, the *order* of the numbers or letters is all-important. Similarly, what the football pools call "permutations" are what we call "combinations"; there is no question of altering the order of the teams within each selection.

Permutational and combinational problems often arise in statistical work, as we shall see as we go on. Card players who wish to adjust their

stakes rationally have to take account of them. What is the probability of
a poker player being dealt a flush in diamonds (i.e., a hand of any five
diamonds?). Here we are not concerned with the *ordering* of the hand, any
five diamonds will do. There are thirteen diamonds in the pack, so the
number of such hands is $^{13}C_5$. On the other hand, the total number of all
possible hands is $^{52}C_5$; so (supposing the pack to have been thoroughly
shuffled) the probability of such a flush in diamonds is

$$P = \frac{^{13}C_5}{^{52}C_5} = \frac{13! \cdot 47! \cdot 5!}{5! \cdot 8! \cdot 52!} \doteq 0.0025†$$

A problem may well have occurred to the thoughtful reader: how
many permutations can be made of objects, some of which are indistinguishable from another? Specifically, if *r* out of *n* objects are exactly alike,
how many discernably different permutations can be made?

Let *x* be the required number. If, in any *one* of the *x* arrangements, the
r indiscernable objects were replaced by *r* different objects, we could form
r! new permutations with these objects, without touching the others. Since
this can be done with all the *x* arrangements, the total number of new
arrangements would be *x . r*!. But this would be the same as the number of
permutations of *n different* objects taken *n* at a time, = *n*!

therefore $x = \dfrac{n!}{r!}$

By the extension of this argument, it follows that if there are *n* objects of
which *q* are of one identical type, *r* of another, *s* of another, and so on,
the total number of discernably different permutations which can be made
is:

$$\frac{n!}{q!r!s!}$$

For example: the number of permutations of all the letters of the word
"greyhounds" is:

     10! = 3,628,800

however, the number of arrangements of the letters of the word
"statistics" (which has three *s*'s, three *t*'s, and two *i*'s) is

$$\frac{10!}{3! \, 3! \, 2!} = 50,400$$

---

† It never happened to me, though I was once dealt a full house, kings on tens, which
is less probable.

In this chapter we have seen how to compute the probability of events when mutually incompatible, and when both mutually incompatible and exhaustive; this was the first law of probability. We have seen how to compute the probability of the concatenation of several compatible but independently uncertain events: this was the second law. We have seen how to compute the probability of an event which depended variously upon other uncertain events: this was the third law. Most important, perhaps, we have derived the theorem of Bayes, whereby we can adjust our estimates of probability in the light of the data we find, in a straightforward and wholly rational manner. Lastly we have seen how many selections and arrangements of given sets of objects can be made. Here, finally, are some problems to exercise and clarify the ideas we have examined.

Exercises

1. A peculiarly fiendish Sultan compels his prisoners to deal three cards from a standard well-shuffled pack. Those who deal three court cards (aces, kings, queens, jacks) must swim across a river swarming with crocodiles; the probability of surviving the swim is 0.2. Those who deal any other hand must walk through a tunnel under the river, the tunnel swarming with cobras; the probability of surviving the walk is 0.01. What is the probability of getting away alive?
2. If someone did survive the attentions of the offensive potentate in question 1, (a) what is the probability that he dealt three court cards? (b) how could you make more sure?
3. There are twenty-four children in a school class. What is the probability that one pair of them have the same birthday?
4. When discussing the screening device for FGR we calculated the probability that a person who was given a positive record was in fact a sufferer. What is the probability that someone with a negative record is in fact free?

References

Lindley, D. *Making Decisions*, Wiley, London, 1971.
Slovic, P. and Lichtenstein, S. "Comparison of Bayesian and regression approaches to the study of information processing and judgment", *Organizational Behavior & Human Performance*, 1971, 6, 649–744.

4

Samples, estimates and other matters

In Chapter 2 we talked about the means, standard deviations and other characteristics of populations. However, in practice it rather rarely happens that we can measure all members of a population for the characteristic that interests us. No man has ever measured the heights of *all* adult male Englishmen, or the reaction times of *all* American females with I.Q.s exceeding 120, or the absolute brightness of *all* the stars in the galaxy. Such total measurements are clearly practical impossibilities. Even on a smaller scale, though it were indeed possible to test to destruction every back axle produced by a factory, such a procedure, though informative, would have the drawback of leaving you with no back axles to sell. In all these cases, we have to be content with *samples*; and from the samples to make *estimates* of the quantities which interest us.

There are about 20,000,000 adult males in Britain. We may suppose that their heights have a mean μ and a standard deviation σ. We have, let us suppose, actually measured the height of 1000 adult males (and, let it be remembered, very few investigations in the Social Sciences use samples as big as that); and the mean $\bar{x}$ and the standard deviation s *of the sample* we now know. How confident can we be of the accuracy of $\bar{x}$ and s as estimators of the unknown quantities μ and σ?

Rather surprisingly, perhaps, we can be very confident indeed; so long as we can suppose that the parent population has a roughly symmetrical distribution about one mode, and so long as σ is finite. In practice, of course, σ may be assumed to be finite: I am unable to think of a practical case where it is not; though I would be intrigued to learn of one. The basis of our confidence is a remarkable theorem known as the Central Limit Theorem. The proof of this theorem requires some rather high-powered mathematics; so it will be omitted here. I shall only refer to it, and discuss what it implies.

One of the consequences of the Central Limit Theorem is that, if some random sample of size n — i.e., of n items — is drawn from any given population of mean μ and finite standard deviation σ and if the mean of such a sample is $\bar{x}$, then the distribution of the variable y defined by

$$y = (\bar{x} - \mu)\, \frac{\sqrt{n}}{\sigma}$$

approaches a Gaussian distribution, of mean 0 and standard deviation 1, ever more closely the larger is n.

This is a bit of a mouthful, so let us decide just what it means. Clearly, from a large population, many samples can be drawn. Reverting to our instance of adult male heights: suppose we took a sample of ten men, we would not be very surprised to find that the mean height of ten men was somewhat different from the mean height of the whole population. If our sample numbered 100 (randomly selected, of course — a point we will have to come back to) we would expect it to be closer; and if it numbered 1000, closer yet. On thinking it over, we would also expect the means of samples of ten to show more variability than the means of samples of 100: after all, since we expect the means of larger samples to cluster more closely about the population mean, they must show less variability than means which need not be so close to it. The Central Limit Theorem tells us not merely that our commonsense intuitions are correct, but precisely how the variability of samples is related to that of the whole population.

It can be proved that $(\bar{x} - \mu)$ tends to zero as n increases; i.e., that the mean of *sample means* approaches μ more closely the larger n is. It can further be proved that the standard deviation of *sample means* is equal to $\sigma/\sqrt{(n)}$; and therefore decreases as n increases.

Thus it can be shown that the *best estimator* of μ is indeed $\bar{x}$. This is not surprising; but we would like to know how close we are likely to be. If we know the value of σ, the properties of the Gaussian distribution (see Chapter 2) would enable us to answer this; and fortunately we can find this value. You will recall that, for a population of N individuals,

$$\sigma = \sqrt{\left[\frac{1}{N} \sum_{i=1}^{N} (\mu - x_i)^2 \right]}$$

It can be shown, though I shall only state the result here, that the best estimator of σ that can be obtained from a sample of size n (where $n < N$) is

$$s = \sqrt{\left[\frac{1}{(n-1)} \sum_{i=1}^{n} (\bar{x} - x_i)^2 \right]}$$

Note that, in calculating the *sample* s.d., we use in the denominator not n but $(n - 1)$. This is because the spread of a sample tends to underestimate the spread of the parent population, as we noted when discussing range in Chapter 2.

Now we have the best estimator for μ and σ; but how reliable is our estimate for σ? Fortunately, we are not fated to pursue an indefinite regress; for it can also be shown that the estimators of σ are themselves approximately Gaussian in their distribution.† Since, as we saw, roughly 68 per cent of a Gaussian distribution lies within ±1 s.d. of its mean, it is usual to accompany a sample estimate with the value of the s.d. of the population of such estimators, which are known from the many consequences of the Central Limit Theorem. These s.d.s are called STANDARD ERRORS — a somewhat misleading term. Thus, when, in a scientific paper we read that a particular quantity has a value of y with a standard error of z, this means that there is roughly a 68 per cent probability that the true value is within the range $y \pm z$. Further, from the properties of the Gaussian distribution, there is roughly a 95 per cent probability that the true value lies within the range $y \pm 2z$, and so on.

(It is fortunate indeed that so many of the properties of samples are roughly Gaussian in their distribution, whatever the distribution of the parent population itself may be. It is hard to imagine what we would do if it were not so; but then, if it were not so, we would be in a different Universe; so speculation is profitless!)

It may help to clarify matters if we state the results we have stated so far, together with one other. These are shown in Table 4.1.

Table 4.1   Estimators

Quantity associated with whole population	Most suitable estimator from sample of size n	Standard error of estimator
μ	$\bar{x}$	$\dfrac{s}{\sqrt{(n)}}$
σ	s	$\dfrac{s}{\sqrt{(2n)}}$
median	sample median	$\dfrac{5s}{4\sqrt{(n)}}$

So far we have talked cheerfully about "random samples" without stopping to consider what this term means. We must now remedy this omission. Again, we enter difficult ground: one of the world's greatest experts on statistics has written that, in some contexts at least, he finds it

† The full story is rather complicated. If the parent population is itself Gaussian, $s^2(n-1)/(\sigma^2)$ is distributed as chi-squared. However, the statements here will not lead the reader into serious error.

"a most baffling conception"·† still, we must try. To do this it is necessary that we talk about sampling *procedure*. Samples, random or otherwise,
are not, as a rule, presented to us on a plate by some superior agency; on
the contrary, we have to go and get our sample. *A sample is called
"random" if it is so selected that every member of the population has a
calculable chance of being selected for the sample.*

Note that the word "calculable" is used, not "equal". This is because,
in some advanced techniques, samples are selected in ways which deliberately weight the odds in favour of some pre-selected scheme. When a
sample is selected so that every member of the population has an *equal*
chance of being selected for the sample, it is called a *simply* random
sample. It is with simply random samples that we will largely be concerned;
and, despite the purists, the word "simply" will be dropped unless there is
some risk of ambiguity.

Our use of the word "random" is, therefore, very different from the
dictionary or "common usage" meaning: "Haphazard, without aim or
purpose or principle, heedlessly." On the contrary, so far from being haphazard or heedless, it requires the utmost care "purpose or principle" to
ensure that a sample is random in the statistical sense.

A very serious problem is that human beings have guessing habits, and
the oddest concealed − and quite unconscious − biases in perception: it is
virtually impossible for a human being to behave randomly. The reader
may try a little experiment: ask as many persons as he has access to if they
would name "at random" any card in a pack. Not only will he find the
court cards to be massively over-represented, but among them, one or two,
such as the ace of spades, will predominate. A far more impressive experiment was reported by Kendall and Babington-Smith (1959). A machine
was devised for presenting truly random digits to a subject, who had
merely to write down what he saw. Table 4.2 gives the digits recorded by

Table 4.2   Digits recorded by a subject using a randomizing device

	0	1	2	3	4	5	6	7	8	9
No. of times recorded	1083	865	1053	884	1057	1007	1081	997	1025	984

one subject. It is evident that this subject had a strong number preference
in favour of even numbers, so that he misperceived on a quite substantial
number of occasions.

It must be emphasized that this subject was not an oddity or any kind

† M. G. Kendall, *Advanced Theory of Statistics*, I.8.3. In the latest edition, co-
authored by Stewart, the term has advanced to "somewhat baffling".

of "nut case": it is quite usual for humans to exhibit this kind of bias. (The reader may possibly have thought that, since experimenters are no more immune from this kind of error than are other mortals, one must treat with reserve a lot of so-called ESP data. If he has thought this, I expect that he is right.)

Thus we cannot rely upon such measures as "I'll pick some subjects at random"; for that can confidently be trusted not to work. What, then, are we to do?

In very simple cases we could toss pennies or throw dice (but don't forget my misadventure with the die in Chapter 1). A more generally useful measure, however, is to use a table of random numbers, such as those given in Appendix Table B. Each digit in this table is an independent sample from a population in which each of the digits from 0 to 9 is equally likely.

The use of such a table is best described by a concrete example. Suppose that we have a population of Army recruits, from whom we require a random sample of ten men. Let us suppose that the population have Army numbers running consecutively from 202,311 to 106,742; i.e., that there are 4432 individuals to choose from. We can take a table of random digits and pick them out four at a time; e.g., from the table open in front of me as I write, I select: 1616, 5704, 8171, 1746, 5329, 7346, 4273, 7763, 6258, 6059 from the top line. We may now take as our sample the individuals having as the last four digits of their service number the groups of four digits from the table.

From the sets I wrote down, we would be unable to use the 1st, 3rd, 4th, 6th and 8th, because these groups lie outside the range 2311–6742; but we have nevertheless selected five of our sample; and we can easily go on to pick the other five.

Of course, few actual populations are as conveniently ready numbered as are service populations; but with a little trouble this method can be extended to cover many – probably most – populations.

Lord Kitchener, who (whatever he looked like in the famous recruiting poster) was nobody's fool, once gloomily observed that you make war, not as you should, but as you must. The same remark might very well be made about experiments in the human sciences. Experiments carried out on a random sample of, let us say, Londoners are rare enough; those carried out on a random sample of human beings are vanishingly few. To confess my own sins, of the thirty-odd experiments I have published, most were performed on Royal Navy sailors, and the rest on Cambridge housewives. Glancing through back numbers of the *Journal of Experimental Psychology* one forms the impression that experimental subjects are most often drawn from the population of university undergraduates. I leave the reader to decide how truly and widely representative such samples are likely to be.

Nevertheless, we must not despair. There are many experimental problems in which there is, at least, no *a priori* reason to suppose that a quite narrowly based sample is likely to be deceptive. For example, in the well-known experiments on the relative memorability of grouped and alternated letters and digits (i.e., on whether it is easier to recall AXP718 or A7X1P8) there is no reason to suppose that undergraduates are different from other people. As always, one does the best one can.

It is often the case that we are only interested in some particular variable or quality; and a sample can be found which is truly random so far as that particular quality is concerned, while far from being so in general. For example, suppose that we are interested in pitch discrimination, there would be no reason to expect that geographical location has anything to do with this, so it might well be acceptable (say) to knock at every twentieth door in a town on a Saturday morning and ask the first person who came to the door. The mention of Saturday morning was not unintentional: there *might* perhaps be a sex difference in this quality, and on weekdays mostly women would be at home.

On the other hand, if we were interested in (say) eye colour, such a procedure might be invalid in an area with a substantial immigrant population; for, at the moment, some immigrant groups tend to be somewhat densely housed and invariably dark eyed, so such a procedure would underestimate the proportion of dark-eyed persons. This merely indicates that the problem of randomization is far from easy, and requires careful consideration.

On the other hand, if, as is often the case, you are saddled with a particular sample and merely wish to investigate the difference between two experimental conditions (a common requirement), it may not be too difficult to randomize. To take a concrete example again: I recently wished to compare the effectiveness of presenting items of information to housewives in two different orders. I am unable to imagine why there should be any systematic relationship between preferences (if any) for ordering information and time of arrival at the laboratory; so I merely alternated the order between successive subjects. Here, as so often in statistics, there is no place for applying rules; rather is there a need for informed judgment.

At this point it is convenient to glance back to Chapter 3 and the problem of the Doctor who was screening patients for Festering Gut-Rot. You will recall that he invented a testing device. The statistical tests we will be looking at in this book are similarly testing devices; though they happen to be numerical rather than chemical or electrical. Like the Doctor's machine, they are liable to two sorts of error — as all testing devices are: they can say "yes" when they should say "no"; and they can say "no" when they should say "yes".

In statistical testing (as, indeed, in all other kinds) we are seeking to distinguish between hypotheses. The Doctor wished to decide between "The patient is free of FGR" and "The patient is suffering from FGR". Behavioural scientists are often interested in deciding between such statements as "There is an important difference between A and B" and "There is no substantial difference between A and B".

Statisticians use technical terms for these possibilities. The hypothesis "There is no substantial difference", or its equivalent (e.g. "The patient is free of FGR") is called the NULL HYPOTHESIS, and is often written H_0. (Read "H-nought.") The other hypothesis (usually the more interesting one, e.g., "The patient suffers from FGR") is called the ALTERNATE hypothesis. In the discussion below we shall denote it by K. The error of *rejecting* H_0 when in fact it is true is called a Type I error; while the error of *accepting* H_0 when in fact it is false is called a Type II error. To save repetition of the phrase "the probability (of some result or value) given H_0" we will use instead the notation $p|H_0$). This is not in the literature, but seems handy.

In any test situation, there will be finite probabilities of making both types of error; and it is not possible to make both of these probabilities arbitrarily small. We will return later to this matter, when we further examine the question of hypothesis testing. For the moment, there is one further point to note. The probability of a Type II error, i.e., $p(\text{accept } H_0 | K)$ is denoted by β (beta, the Greek small b); and the POWER of a test is defined as:

$$\text{power} = 1 - \beta$$

In other words, the less likely you are to make a Type II error, i.e., the less likely you are to accept the null hypothesis H_0 *in error*, the greater is the power of your test. This is quite reasonable, when you think about it.

In this short chapter we have seen that, thanks to a fortunate theorem, measures obtained from samples can be referred to a whole parent population. We have further seen that the "standard error" enables us to establish confidence limits for these measures. There are, however, real difficulties in ensuring that the samples we use are truly random; and, although such devices as tables of random numbers are a great help, there is no substitute for informed good sense in finding a random sample.

We have briefly considered the two types of error to which all statistical tests are liable, and have introduced and defined the concept of the "power" of a test. Here are some exercises to underline these points.

Exercises

1. In Chapter 2 we treated the children whose scores are given in Table 2.1

as a whole population. Treating them now as a sample, estimate the mean and s.d. of the parent population, and comment on the reliability of these measures.

2. A newspaper polls its readers and announces that 60 per cent of the country is opposed to X. Comment on this statement.

3. A test has been devised to measure the viciousness of schoolchildren. An education authority which runs ten secondary schools, each with an annual intake of about ninety, wishes to have an idea of the viciousness of the thirteen-year-old children in its care. Suggest a suitable sampling procedure, if there is reason to think that 30 is a suitable value for n.

4. The displeasing Sultan we encountered in Chapter 3 is given to setting statistical problems for his Viziers. They are free to choose their own techniques, but if any concludes that K is false when it isn't he is fined 50 dinars, while any who concludes that K is true when it isn't is cast into a pit with a collection of poisonous reptiles. The Vizier Isaac O'Brien Ibn Mohammed plumps for the most powerful test in the book. Is he wise in his generation?

5

What's the problem?

Marshall Foch, who seems to have been weak at statistics, used to cite with great approval the story of a general who arrived on a battlefield and said, "To hell with all the theories, *what's the problem*?" That general might have made a good statistician; for his question was the first and most vital one for any applied statistician to ask. There is no use in, nothing indeed is more futile than, merely "gathering data" and hoping that enlightenment will emerge after suitable fiddling with the numbers. The first thing one has to have clear is: what is the problem? *What do you want to know?*

Sometimes the question asked is the basic and simplest one: what is the population *x* like? This question can be answered, as we saw in Chapter 2, by measures of location (mean, median, mode) and dispersion (s.d., range). More usually, in the behavioural and social sciences, we want information about *differences* or *associations* between different, or apparently different, populations. These matters require elaboration.

The Bear, as you will recall, went over the mountain to see what he could see. This is a perfectly good reason for going over mountains; but it might have been useful if he had started with some more specific idea in mind – e.g., was there more honey per square kilometre on the other side; or even, very basically, was there any difference in the honey density on the two sides of the mountain?

It is the latter case that is more typical of most statistical investigations. We start off with some idea about the world: more impressively, an hypothesis. This idea may be sharp and unambiguous; e.g., we might have the idea that women will make fewer errors in a car-driving task than will men. The idea may be more general: e.g., we might merely suggest that performance of men and women in a driving task will be different. The idea may be one of a continuous relation; e.g., we might suggest that errors in driving will be fewer the longer the subject has been driving. We are not, at the moment, asserting that any one of these statements is true; they are here merely examples of the kind of idea which can lead to useful investigations.

Each of these kinds of questions can be approached in a variety of ways, as we shall see in succeeding chapters; and for each approach there is an appropriate statistical technique. But again, what can we get from these techniques? As we saw in Chapter 3, we can not get a *certain* answer: we can never assert that it is absolutely certain that, e.g., women are genuinely different from men as drivers. We can, however, be sure *enough* to decide, for example, that insurance premiums can safely be different. What we are looking for then, is, quite generally, *sufficient grounds for action.*

"Action" here does not necessarily mean more than deciding to accept a particular theory for the time being. It could, on the other hand, mean a demand or a decision for far-reaching legislation. Clearly, what is "sufficient ground" depends on the sort of decision to be made; and this logically brings us to the discussion of an important and highly deceptive term: the word SIGNIFICANCE.

It is very unfortunate that a technical term is also a word in common usage, especially when that common usage is notably different from the technical one. The reader will recall that, however good the agreement, or however sharp the disagreement between observation and theory, there always remains some finite probability (which may be negligibly small) that this agreement or contradiction arises from sheer chance. For example, the evidence which lead me to reject the use of the die mentioned in Chapter 1 could have arisen by purely chance happening; but this was so improbable that I decided on the rejection. Now, what level of probability are we going to regard as negligible? Or, to put it the other way round, what level of evidence will we require in order to act? Clearly this depends on the *consequences* of our action: there is no fixed level of evidence which is always adequate.

Suppose we are conducting a preliminary investigation of some interesting field concerning which we have some provisional hypothesis in mind. We conduct a pilot experiment, and the results seem to support our notion. If we are right, we shall proceed to carry out a further, more rigorous test, which would be a waste of time if we were wrong. However, even so, no great harm would be done; and we might very well decide to proceed to the second stage so long as the probability of our being in error were (say) less than 0.1. Suppose, on the other hand, that we were asked to advise on the suitability of some suggested procedure for reducing road accidents — a procedure which, let us say, looks promising but expensive. Suppose we were wrong in our advice, this might result in the waste of millions of pounds and the loss of many lives. We would want to be very confident in this case; and we might very well demand that the probability of our results being mistaken should be (say) less than 0.001 before deciding.

In both these cases, the acceptable probability of being wrong is called the *level of significance* required. It is sometimes denoted by the Greek letter α; and in the first case we would put $\alpha < 0.1$, and in the second <0.001 (α is the small a, alpha).

There is a convention — it is nothing more than that — to refer to any result which satisfies an α of <0.05 as "significant", and any which satisfies an α of <0.01 as "highly significant". This is fair enough, as a rule of thumb; but it has had some unfortunate consequences. First, and less importantly, many tables of statistical functions are printed for only these values. Sometimes it is easy to interpolate for the value you want, but not always. This imposes some limitations on your freedom of choice; but it is not otherwise too serious.

Much more important is the fact that the values 0.05 and 0.01 have acquired "magic number" status, and that some of the common usage meaning of "significant" has rubbed off on them.

Many editors will not print results which fail to meet the criterion $\alpha < 0.05$; and many authors, if they mention such results at all, do so with a deprecatory phrase, such as "the findings were not quite significant". But this is merely silly. It is true enough that a finding which appears to upset some well-established notion, or one upon which expensive or possibly dangerous action is to be based should not be accepted without a really small α. But a result which is not, and which its finder does not pretend to be, more than a general pointer towards further and more precise work should surely be considered at an α of (say) 0.1. Further, it should not be forgotten that it is the job of the investigator to set α before carrying out his study; it is not a general value to be set arbitrarily for all studies. Treating 0.05 as a magic number has made many workers un-willing, or unable, to publish interesting results which might be well worth looking at; and has made others go into absurd and intellectually dubious arithmetical contortions in order to shove some set of data over the borderline. I do not know what should be done about this, except to keep pegging away with a sensible approach.

Treating "significant" in its technical sense as synonymous with "significant" in its common usage is subtly pernicious. A finding which is technically highly significant *may* be extremely significant in the usual sense; but equally it may not. In common parlance a "significant" factor is one which is important in some practical situation; but in the technical sense one can obtain highly significant results and only be confident that the factor is operating, importantly or otherwise. For a competent experimenter will, for a start, scrub out all factors extraneous to the one he is studying. If he does this successfully, and if the factor he has elected to study has any real effect at all, he may obtain results which are *technically* significant at a very rigorous level indeed. Whether that factor is commonly "significant" — i.e., whether it plays a large part in a practical situation — may require further and quite separate study.

Let us consider an admittedly rather far-fetched example. Someone decides to see whether there is a gravitational attraction between two half-million-ton supertankers 1 km apart. He makes sure the engines are not running, the sea is still, there is no wind, the magnetic fields of the hulls are de-gaussed and so on; and then sets up his Oetvoes balances and lo! his results differ "significantly" from zero. Indeed, since the force is (if I have done my sums aright) something under 1 kg weight, his results should be massively "significant". I leave the reader to guess how "significant" they would be to a couple of moving supertankers in the area of harbour mouth.

I do not suggest, of course, that the term "highly significant" has been applied to behavioural or social experiments quite as absurd as the Case of the Gravitating Supertankers: but the subtle deception is often — quite unconsciously — used; and we must be on our guard against it.

Back to the sorts of problems we must cope with.

When we have analysed our problem, decided upon our measure, carried out our tests and collected our data, we almost always find ourselves left with one of two questions to answer: is A greater than B? or is A different from B? (B can, of course, be zero). A and B will be some parameter (usually, though not always, the mean or median) associated with batches of data. The former is called *a one-tailed question* and the latter *a two-tailed question*: they are respectively resolved by one- or two-tailed tests. These terms we must explain.

As we saw in Chapter 4, the fortunate Central Limit Theorem showed us that, in general, over a wide range of forms of the parent population, the mean or median of samples will themselves have an approximately Gaussian distribution. The population of values of the estimators of B will then have a Gaussian distribution as represented in Fig. 5.1. If we are asking "Is $A > B$"? and setting — let us say — an α of 0.05, we are asking: is the value of $A > 95$ per cent of the values of the distribution of Bs.

In other words (see Fig. 5.1a), to be satisfied, we are demanding that the value of A lie in the right-hand 5 per cent tail of the distribution of Bs.

In the other case, where we merely ask "Is $A \neq B$?", we are not pre-electing the tail of the B distribution in which A is to be acceptable: either will do. However, if we set the same value of α, 0.05, this acceptable fraction must now be split between the two possible tails wherein A may be found. Thus (see Fig. 5.1b) to satisfy our criterion in a two-tailed case A must lie in either of the $2\frac{1}{2}$ per cent tails of the B-distribution.

It is important to be clear about the meaning of one- or two-tailed questions, precisely because quite mistaken notions on the subject have

Fig. 5.1   One- and two-tailed acceptance zones for alternative hypothesis

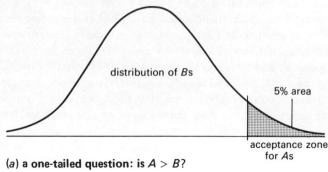

(a) **a one-tailed question: is $A > B$?**

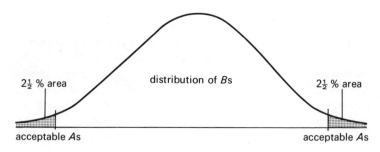

(b) **a two-tailed question: is $A \neq B$?**

very wide circulation among behavioural scientists. Many speak or write as if a two-tailed test were, in some mystical way, superior to a one-tailed one; and one often sees in papers which report the testing of unambiguous hypotheses of the $A > B$ type the solemn phrase "The test was significant (two tailed)". (*Peccavi!*) This is absurd. It is *the question being asked* which decides whether one- or two-tailed tests are *appropriate* (not "best") and this should never be forgotten.

Discussing our problems above, I used the words "When we have . . . decided upon our measure", and I expect that the reader glided over that phrase without much worry. Alas, it is not always a trouble-free decision. Unfortunately, no general rule or guidance can be given for selecting the appropriate measure in a given case, other than the universal one: let us use our commonsense.

Sometimes it is easy. When, in Chapter 1, our Cetian friend wanted to investigate sexual dimorphism in humans he used the immediately obvious measure of *height*. When examining the effect of prior warning upon response facilitation the measure is again obvious: reaction time (in milliseconds). Other cases are less easy. How should we compare the effectiveness of two methods of teaching children to read? Clearly by comparing performance when reading some selected passage which they had not previously seen. But what is to be the measure? The time taken? The number of omissions? The number of errors? The number of correct answers to a set of questions designed to see whether the passage was understood? Some combination of these?

No doubt the problem is soluble, but you will see that there is no immediately obvious and unquestionable measure. In some cases, indeed, the difficulty of finding a measure has lead some people to regard questions concerning which there are masses of good numerical data as, nevertheless, unanswerable. Let us examine a case which is of no great moment, in itself, but which well exemplifies the kind of problem one comes up against.

In the pages of Wisden, and other works, we can find masses of data concerning the performances of many great cricketers; but enthusiasts for the game (among whom I am happy to be included) are often at odds when attempting to rank-order the players. Suppose we limit ourselves to batsmen only, you may hear such remarks as "X was a fine bat, but in his era the pitches were so much better than in Y's day that he was bound to have a better average; though he was no better, if as good", or "Y was damned good, but how would we have coped with bowler Z?"

Such questions are not frivolous; and indicate the kind of real difficulty a statistician can come up against; the "obvious" measure — in this case, the batsman's average — has sufficiently grave limitations to be of questionable value. Nevertheless, given thought, one can usually find a measure. I will suggest one for this case which the reader can amuse himself by picking to pieces.

What's the problem? Clearly to measure *outstandingness*. We may suppose that a player with a reasonably long career would have been as outstanding in any era other than his own, even though his actual scores would doubtless have been different. My suggested solution, then, is this: consider test careers only (when the opposition was strongest), and use as your measure the proportion of the total runs made by X's side which he himself made, in the matches in which he played.

In more serious questions than this sporting one, analogous difficulties arise: how, for example, do you compare the social status of two individuals, or of two professions? Even tougher: how do you measure relative change in status over time? Without going any further, the reader will see how the problem of selecting measures can range from the trivially easy to the exceedingly difficult.

In this chapter we have emphasized the importance of clearing the ground before starting any statistical investigation. Above all, the investigator must be absolutely clear about *what he wants to know*. He must then decide, in the light of the consequences of his decision, what level of evidence will satisfy him; he may have to devise a suitable measure for the object of his interest; and he may then proceed to a fruitful conclusion. If these matters are left vague, the result is likely to be a botch.

Exercises

1. A particular disease is always fatal if untreated, but a cure rate of 70 per cent is achieved using a known therapy "A". A different treatment ("B") is suggested: comment on the problem.

2. It is suggested that the spread of TV is a cause of the increase in crime. Comment on the problem.

3. A certain pharmaceutical product is guaranteed by its manufacturers to be 96 per cent pure, and they offer a substantial cash indemnity if any item sold fails to meet this criterion. They can be sure that the mean purity of this product is 98 per cent, and employ a statistician to see that they keep their money. Comment on his problem.

6

A binomial tale

It often happens in the behavioural and social sciences that the data we are examining fall naturally into two classes or sets. Those who have been involved in motor accidents might be classified as dead or alive; applicants for certain classes of jobs as men or women; the answers to some cognitive task as correct or incorrect. More often, perhaps, it is still possible to take a basically continuous distribution and dichotomize it in some reasonable fashion: e.g., survivors of motor accidents could be classified as seriously injured or not, a possible dividing line being retention overnight in hospital.

Again, it is often the case that we have some *a priori* reason to expect a particular balance of probabilities, and have but a single datum against which to test this expectation. To pursue the examples given above, we might expect there to be no sex difference in a particular job situation (i.e., the probability of an applicant being male = the probability of an applicant being female) and find that eight out of twelve applicants were female. Do we decide that our expectation looks as though it had been justified or not?

In cases like this, we can often use a Binomial test; which is so called because it is based on the Binomial Theorem. As you may well guess, this test is very widely suited to many questions in the behavioural and social sciences; and when it is properly applicable, it can be astronomically powerful. In my opinion, it is not used in practice as often as it might be, since many workers in the field show a quite inexplicable preference for techniques which are more fancy without always being more effective. We will now state and prove the basic theorem; and can assure our readers that neither statement nor proof is really difficult, though they may look a little unfriendly.

Consider these very simple algebraic expansions:

$$(1 + x)^2 = 1 + 2x + x^2$$
$$(1 + x)^3 = 1 + 3x + 3x^2 + x^3$$
$$(1 + x)^4 = 1 + 4x + 6x^2 + 4x^3 + x^4$$

We could go on; but stopping for a moment we notice that each expansion
contains all powers of x up to and including the power to which (1 + x) is
raised. We see that the first and last coefficients are always 1; and we might
notice (but probably won't unless someone points it out) that the values
of the coefficients are related in an interesting way to the combination
values which we talked about in Chapter 3. In fact we can re-write these
expressions:

$$(1 + x)^2 = 1 + {}^2C_1 x + x^2$$
$$(1 + x)^3 = 1 + {}^3C_1 x + {}^3C_2 x^2 + x^3$$
$$(1 + x)^4 = 1 + {}^4C_1 x + {}^4C_2 x^2 + {}^4C_3 x^3 + x^4$$

which suggests the possibility that for any positive integer:

$$(1 + x)^n = 1 + {}^nC_1 x + {}^nC_2 x^2 + \cdots + {}^nC_r x^r + \cdots + x^n \ldots \qquad (6.1)$$

That this is so can be simply proved. Let us *assume* that (6.1) is true for
some particular n; and multiply it by (1 + x). Then, if we put all terms in
the same power of x together, we will obtain:

$$(1 + x)(1 + x)^n = (1 + x)^{n+1} = 1 + (1 + {}^nC_1)x + ({}^nC_1 + {}^nC_2)x^2 + \cdots$$
$$+ ({}^nC_{r-1} + {}^nC_r)x^r + \cdots + x^{n+1} \qquad (6.2)$$

but
$$^{n+1}C_r = {}^nC_{r-1} + {}^nC_r$$

therefore

$$(1 + x)^{n+1} = 1 + {}^{n+1}C_1 x + {}^{n+1}C_2 x^2 + \cdots + {}^{n+1}C_r x^r + \cdots + x^{n+1}$$
$$(6.3)$$

Thus we have shown that, *if* (6.1) holds true for some n, it is also true for
$n + 1$. But we have already seen that it is true for $n = 2$, $n = 3$ and $n = 4$.
Therefore it is true for $n = 5$; therefore . . . and so on for ever: i.e., it is
always true. (This kind of proof is properly called "proof by recurrence
relation". For reasons wholly mysterious to me it is often called "proof by
mathematical induction": the method is simply not inductive.)

By exactly similar means, which the reader may try for himself, we can
prove the more general case:

$$(x + y)^n = x^n + {}^nC_1 x^{n-1}y + {}^nC_2 x^{n-2}y^2 + \cdots + {}^nC_r x^{n-r}y^r + \cdots + y^n$$
$$(6.4)$$

This is the *Binomial Theorem*.

To save the trouble of calculating the several ^nC_r values they can simply
be written down using the following procedure. Write down the coefficients of $(x + y)$, i.e., 1 ,1. Below the first 1, write another; below the

second, write the sum of the number above it, and of the number above it and to the left, i.e., 1 + 1 = 2. The vacant space to the right is filled in the same way, i.e. with 0 + 1 = 1. Fill in the third row in the same way, obtaining 1, 3, 3, 1, and so continue. Thus we obtain the array of numbers given in Fig. 6.1, which is known as Pascal's Triangle. (Named after Blaise Pascal, a French mathematician of the seventeenth century, who, as you may recall, was terrified by the eternal silence of infinite space, quarrelled with the Jesuits and found that mathematics cured toothache.) The nth row of this array gives the coefficients of the several terms in the expansion of $(x + y)^n$, for all n.

Fig. 6.1   Pascal's Triangle

1	1					
1	2	1				
1	3	3	1			
1	4	6	4	1		
1	5	10	10	5	1	
1	6	15	20	15	6	1
—	—	—	—	—	—	—

and so on

The reader may well have decided by now that this is all very interesting, but, despite the appearance of combinatorial terms, be wondering how we go from here to applied statistics. We are now coming to this directly; and the ^nC_r terms indeed provide the clue.

Consider any two *mutually exclusive and exhaustive* events A and B. Let the probability of A be p, then that of B must be $q = (1 - p)$. Let the *combined* outcome of two events be observed. It will be immediately obvious that we can have just three observations; two As, two Bs, or an A and a B. From Chapter 3, we can easily see that the respective probabilities of these observations are p^2, q^2 and $2pq$ (and, naturally, since one of them *must* occur, the sum of these three probabilities is 1. The reader can easily prove this, substituting $q = (1 - p)$).

Quite generally, if we are looking at the *combined* outcome of n trials, there are $n + 1$ possible results, namely, n As, $(n - 1)A$s and $1B$, $(n - 2)A$s and $2B$s, and so on to n Bs. The respective probabilities of the $n + 1$ possible outcomes are shown in Table 6.1.

Table 6.1   Probabilities of all possible outcomes of n trials

Outcome	n As	$n-1$ As, B	$n-2$ As, 2Bs	... $n-r$ As, r Bs	... n Bs
Probability	p^n	$^nC_1 p^{n-1}q$	$^nC_2 p^{n-2}q^2$	... $^nC_r p^{n-r}q^r$	... q^n

If this looks familiar, it is because all the terms in the second row of the table are formally identical to the terms of (6.4): they are the terms of the Binomial expansion. Also, just as in the three-alternative case we discussed above, the probabilities of the several outcomes summed to 1, so, in the general case:

$$\sum_{r=0}^{n} {}^nC_r p^{n-r}q_r = 1$$

Thus, in any case where the observations fall naturally into two classes, or where they can reasonably be divided into two classes, we can use the Binomial Theorem to find either or both of two things: we can find the probability of a *particular outcome*, and (often more important) we can find the probability of any outcome *as likely as*, *or less likely than*, *any particular outcome*. The difference — which is an important one — between these can best be brought out by considering an example.

A fairish marksman hits a given target 80 per cent of the time. Thus the probability p of a hit (H) is 0.8; and hence the probability q of a miss (M) is 0.2. He fires a clip of 5 rounds at the target. What is the probability that (i) he misses every time? (ii) he misses once? (iii) he misses *at least* once? (iv) he misses *at least* twice?

Here $n = 5$, and the 5th line of Pascal's Triangle gives us the required coefficients, so the only arithmetic we do is to compute the various powers of 0.8 and 0.2. This done, we can easily tabulate all the possible outcomes, and their associated possibilities (Table 6.2). (A useful check of

Table 6.2   Probabilities of outcomes of a shoot

Result	Probability
5H	$p^5 = \quad 0.8^5 \quad = 0.32768$
4H, M	$5p^4q = 5 \cdot 0.8^4 \cdot 0.2 = 0.40960$
3H, 2M	$10p^3q^2 = 10 \cdot 0.8^3 \cdot 0.2^2 = 0.20480$
2H, 3M	$10p^2q^3 = 10 \cdot 0.8^2 \cdot 0.2^3 = 0.05120$
H, 4M	$5p \, q^4 = 5 \cdot 0.8 \cdot 0.2^4 = 0.00640$
5M	$q^5 = \quad 0.2^5 \quad = 0.00032$

your arithmetic is to ensure that the sum of the several probabilities is exactly 1.) We can now answer our questions. The probability of 5M is

0.00032, which is pretty small. The probability of 1 miss, i.e. the probability of 4H, M is, as we see, 0.4096. The probability of *at least* 1 miss, however, is the total probability of all outcomes other than 5H, i.e., it is $1 - 0.32768 = 0.67232$. Similarly, the probability of at least 2 misses is the total of the probability if 2, 3, 4 or all 5 misses, which = 0.26272.

Note especially the difference between the probability of *one* miss and of *at least* one miss: in this particular case the difference between a less-than-evens and a better-than-two-thirds chance.

A commonly occurring case is one where we wish to test the hypothesis that $p = q = 0.5$. A typical instance is the one we imagined earlier: our expectation was that equal numbers of men and women would apply for a post; but out of twelve applicants only four were men. This is not quite as dead simple as it looks; so let us consider it in some detail.

In the case $p = q = \frac{1}{2}$, for any value of n, $p^r q^{n-r} = \frac{1}{2^n}$. Consequently we can write down the probabilities associated with each outcome by writing down the appropriate line from Pascal's Triangle — the nth line — and dividing each term by 2^n. Thus, if $n = 3$, the possible outcomes, calling each item an A or a B, and their associated probabilities are: $3A$s, $\frac{1}{8}$; $2A$s$1B$, $\frac{3}{8}$; A, $2B$s, $\frac{3}{8}$; $3B$s, $\frac{1}{8}$. Since $2^{12} = 4096$ we can readily tabulate possible outcomes of twelve job applications and their associated probabilities (Table 6.3).

Table 6.3 Probabilities of applications for a job: $p = q = \frac{1}{2}$

No. of male applicants	0	1	2	3	4	5	6	7	8	9	10	11	12
No. of female applicants	12	11	10	9	8	7	6	5	4	3	2	1	0
Probability ×4096	1	12	66	220	495	792	924	792	495	220	66	12	1

It is worth taking a moment to look at Table 6.3. The most probable outcome, evidently, is $A = B = 6$; but its probability is somewhat less than $\frac{1}{4}$: it is *most likely* but NOT *"more likely than not"*. There is, however, a better-than-even chance of 5 *or* 6 *or* 7 of either sex; the probability of one of these being the outcome is roughly $\frac{5}{8}$.

Now it is important to be quite clear what question we are asking. Are we asking "Is $p = q$?"; or are we asking "Is $p > q$?". Let us take these two in order.

If we have doubts merely as to whether $p = q$, we must ask: what is the probability of there being *as few as* 4 (i.e. any number from 0 to 4) of *either* sex? This is a two-tailed question: we are as interested in extreme frequencies of men as of women. What sort of α should we pick? This is not, *a priori*, a vital question, so we might decide to do further work if $P \leqslant \frac{1}{10}$. In fact P is much higher than this: the table shows that it is $\frac{1488}{4096} \doteq \frac{3}{8}$. So we do not chase this hare any further.

The alternative question is one-tailed: we ask for the probability of there being as few as four *men* out of the twelve. The α we have already selected still seems reasonable: and the table again gives P. It is $\frac{744}{4096}$ or roughly $\frac{3}{16}$: again too large to justify further effort.

The reader will have noted how simple the binomial test becomes when $p = q$. Indeed, it is hard to think of a simpler test for this hypothesis; for if we set α in advance, it is a simple matter to write down the largest number of As (As being the least frequent outcome) which would satisfy that α, either for one- or two-tailed questions, for any given n. For example, it is evident from the table that if we decide that $\alpha = 0.05$, and our question is two-tailed then an A of 0, 1 or 2 will satisfy α; i.e., 2 is the upper bound for a two-tailed test on an n of 12 and an α of 0.05. Naturally, it follows that 2 is the upper bound for a one-tailed test on an n of 12 and an α of 0.025. To save even the slight labour of working out the successive lines of Pascal's Triangle, Appendix Table C gives the one-tailed probabilities associated with As as few as a series of given numbers for a set of ns. (It will be seen from the table that 4 or fewer in an n of 12 has a P of 0.194; not far from $\frac{3}{16}$ as we roughly gave.) The two-tailed Ps are obtained by simply doubling each value in the table.

As the reader will see, the table only covers values of n up to 25. Beyond this, using the theorem can be a trifle tedious (though trivially easy if you have access to a computer). Fortunately, there exists a convenient approximation which can be used to circumvent excessive arithmetic.

If the reader sketches a histogram whose heights are the values of Table 6.3, he will immediately notice that the outline looks very like a stepped version of the Gaussian curve. It may occasion no surprise, then, to learn that, as n becomes large, the binomial distribution approximates closely to the Gaussian; and in the limit, as $n \to \infty$ is indistinguishable from it. This fact can be exploited for large n by using a Gaussian approximation. The reader will have noted that, whatever n is, the most likely result is np, or the two values nearest to this for odd n. Suppose that the observed result is x. Then (recall the Central Limit Theorem) it can be shown, although the proof is rather high-powered, and will be omitted, that the distribution of x is approximately Gaussian with $\mu = np$ and $\sigma = \sqrt{(npq)}$ which gives, if $p = q = \frac{1}{2}$:

$$\mu = \frac{n}{2}; \quad \sigma = \frac{n}{4}$$

Thus, we could use the unit normal table (Appendix Table A) for large n by using a new variable z obtained by the simple transformation

$$z = \frac{x - \mu}{\sigma}$$

i.e. $\quad z = \frac{|x - np|}{\sqrt{(npq)}}$ \hfill (6.6)

There are, however, certain snags for all finite n — i.e., for all real problems — which may well have occurred to the reader, and for which corrections need to be applied. These snags are that the binomial distribution, unlike the Gaussian, is discontinuous; and, that, if $p \neq q$, the binomial distribution is asymmetrical.

Once again, I shall not give the derivation of the corrections and precautions which have to be used, but shall merely state them. They are:

(i) Never use a Gaussian approximation if $n < 25$.

(ii) Reduce $|x - np|$ by 0.5, i.e., alter (6.6) to read:

$$z = \frac{|x - np| - 0.5}{(npq)}$$ \hfill (6.7)

(iii) If $p \neq q$, only use the Gaussian approximation if $npq \geq 9$.

As always, these rules and procedures are best illustrated by an example.

Let us suppose that an animal experimenter has noticed that, out of twenty-five occasions when he put untrained rats into a T-maze, the beasts turned left on twenty of them. He wonders whether this shows a true preference. (Not necessarily, by the way, a handedness preference. The rats' sensory world is somewhat different from ours: there might, e.g., be a smell emanating from the left-hand end of our laboratory, which is attractive to rats, but which we cannot detect.)

Obviously he asks a one-tailed question: is p (left) $> p$ (right)?; and he wishes to know the probability of his observed result ($x = 20$), or of any result further from equal numbers of left and right turns, arising simply from $p = q$. He is prepared, let us say, tentatively to accept a genuine difference if the probability given by $p = q$ is $< \frac{1}{50}$; i.e., his α is 0.02.

The value of np is clearly 12.5, and of $\sqrt{(npq)}$ is 2.5. Substituting in (6.7) for the value of deviation to look up in Appendix Table A, then, gives:

$$z = \frac{|20 - 12.5| - 0.5}{2.5} = \frac{7}{2.5} = 2.8$$

The table shows that the probability of a result as extreme as this is

$(1 - 0.997) = \frac{3}{1000}$. This is well within the chosen α; and our experimenter decides that he is on to something. Exactly what he is on to is, of course, for him to find out. A statistical test does not do his thinking for him.

So far in this chapter we have been looking at cases where there is effectively a single datum: eight applicants out of twelve were women; on twenty occasions out of twenty-five the rats turned left; and so on. The binomial test, however, is so simple and powerful, and so relatively undemanding that one might well ask whether it has other applications. Indeed it has. A word of caution is necessary, however, for no test is completely undemanding. It is assumed that the *probabilities p and q are constant during the time of observations*, and that *the several trials are truly independent*. The user must be satisfied that these requirements are met. We can now proceed to look at another case in which the binomial test can be used.

Two sample problems

First a jargon word must be defined. This word is "TREATMENT". In statistics a "treatment" is that variable which interests us. We may want to know which of several methods of education is best — or whether they truly differ in their effects at all. We may wish to know which of two fertilizers is preferable for encouraging tomatoes; which of two advertising campaigns is most effective; which of various drugs best arrests the progress of some disease, or which dietary item induces it. In all these cases, the interesting variable — the teaching method, the fertilizer, the advertising campaign, the drug or the menu — is called the "treatment".

Let us limit ourselves for the moment to discussing the case when two, and only two, treatments are to be compared. It is obviously necessary that the two groups of subjects upon whom the treatments are to be tested must be comparable — a requirement easier to state than to satisfy. As a general rule, experimenters take some tolerably homogenous group, such as first-year science students, and divide the group in twain by some procedure which they hope is random, and which in any event should not favour any relevant variable.

More ambitious workers, however, may attempt to match each individual in one group with a specific member of another. Though attractive if you can do it, this is extremely difficult. Suppose you wish to match Mr X from group 1 with Mr Y from group 2. You may know that they are the same age, in equally good health, within three points of the same I.Q., and of equal educational attainment: but can you, at this stage in the development of the behavioural sciences, be sure that you have matched them for all the relevant variables? After all, there is an eloquent school of thought which maintains that a factor of great importance is whether or not their mothers were always within 50 m of them for the first two years of their lives. This may sound idiotic — it very likely is — but we cannot yet rule it out absolutely; and it is very hard to find out the status in such matters of given pairs of individuals.

In an endeavour to outflank all these difficulties, a frequent variant of the matched-pair technique is to make each individual his own control. The procedure is to halve a group of subjects in some random way, and to give one half the two treatments in succession in the order 1, 2 and to give the other half the two treatments in the order 2, 1. This certainly avoids the difficulties of matching different individuals; and it is hoped that, by taking both orders of presentation, it obviates unwanted transfer effects.

The difficulty is that, if you give an individual two treatments, the second one is not given to the identical person, but to a person *who has received the first treatment*. With human subjects especially, this can land us in the dark waters of "transfer of training" (see e.g., Hammerton, 1967). The purpose of dividing the experimental population in two and reversing the order of presentation in each half is to minimize this unwanted transfer effect. It may not be completely successful in this — a matter we shall have more to say about later — but it is the best there is. (See, e.g., Poulton and Freeman, 1966.)

Thus, when a matched-pairs technique is used, we characteristically end up with a set of pairs of readings, which have been obtained either from matched individuals, or from the same individuals on separate occasions. When and how can binomial tests be applied to such data? The answer to this question is that they always *can* be used, but they are best used when the two treatments are merely ranked, not measured. Thus, when the result of treatment 1 is graded as merely less than, equal to, or better than the result of treatment 2, the binomial test is probably best. When actual scores are available (e.g., the subject learned ninety-five words in an hour using method A, but only seventy-six using method B) other methods can be used, which can extract more from the data. Let us consider an imaginary case.

A car designer wishes to compare two steering-wheel positions for a new car. Let us suppose that one of these positions (1) is that adopted in all previous models, while the other (2) is a novel one which, it is hoped, will offer certain advantages. Two prototype versions of the car are available, one with 1, the other with 2. A panel of twenty drivers has volunteered to test the two versions, ten are to have 1 then 2, the other ten in the reverse order.

Clearly, this is a one-tailed question: is 2 better than 1? It were possible to construct performance measures, such as number of movements made during a prescribed course, which would enable a number to be attached to each driver's performance with each type of wheel. However, in the interests of speed (of experiment, not of the vehicles) each driver was merely asked whether either appeared better — i.e., easier to drive — or whether there was no difference. An opinion that 2 was better than 1 was recorded as "+"; that 1 was better than 2 was recorded as "−"; and no detectable difference was recorded as "=". Table 6.4 gives the results obtained.

Table 6.4   Opinions of twenty drivers comparing two steering-wheel positions

Driver No.	Opinion	Driver No.	Opinion
1	+	11	+
2	=	12	+
3	+	13	=
4	+	14	+
5	+	15	=
6	−	16	+
7	+	17	+
8	+	18	−
9	+	19	+
10	−	20	+

The α in this case is set largely by economic considerations. The alteration to the production-line jigs which would be demanded by a change to 2 would be large, though not insurmountably so. We will suppose then that the production side insisted on an α of 0.01.

The value of n here is 20; but the reader will note that three judges found themselves unable to distinguish between the two positions. Such indistinguishable outcomes are referred to as "ties" or "tied scores". What should be done about them?

The answer to this question depends on the test being used. In the case of a binomial, ignore them, and proceed as if n were reduced by the number of ties. In this case, then $n = 17$, and we wish to know the probability of a distribution of results as unlikely as, or more unlikely than 14 +s and 3 −s, supposing that there is indeed no difference between the two positions. (This, of course, is an equivalent statement of the probability of being wrong in supposing that the two probabilities are indeed different in the supposed direction.) Appendix Table C gives $p = 0.006$. This is <0.01 and we therefore recommend to the production branch that the new form be adopted.

In this example, we have used readings which effectively take only two values (+ and −, since = is ignored) but clearly there is nothing to prevent us using it for any matched-pair situation: *any* two scores can be rated as >, < or =. Thus for *all* matched-pairs data, the binomial test *can* be used. (In my judgment it should always be used, at least as a first step.) It has a disadvantage, of course, when readings take the form of numbers. This disadvantage is that it takes account only of the *direction* of differences, not of their magnitude. Clearly there is loss of information here; and there are more refined methods, which we shall look at later, which make use of this extra information.

We have seen in this chapter that the Binomial Theorem provides a neat and effective tool for coping with problems which arise when the data can readily be divided into two classes: A and $\overline{A}$, whatever their underlying distribution may be, so long as the several readings are independent, and the associated probabilities constant. It can be used both to test whether $p(A) = p(\overline{A})$ and to test some specific hypothesis of the form $p(A) = p$, $p(\overline{A}) = q$. It can be used both in single observations and matched-pair situations. To conclude we will outline the procedures to be adopted in these several cases.

In all cases

(1) Decide whether you have prior information leading you to test whether $p(A) = p$, $p(\overline{A}) = q$, or whether you are simply inquiring whether $p = q$. (The latter is by far the most common occurrence.)

(2) Decide whether you are asking whether $p = q$ (two-tailed problem) or specifically whether $p > q$ (one-tailed). Select as appropriate in the light of general circumstances. Then proceed as follows:

Single observation (*a*) Out of a sample of size n ($n < 25$) x have characteristic A, the rest $\overline{A}$. (*i*) Testing $p = q = \frac{1}{2}$. *Either* use the nth row of Pascal's Triangle, dividing each item by 2^n, to obtain a table of probabilities, and find the probability of the observed outcome, or any more extreme. *Or,* use Appendix Table C to save the trouble.

(*b*) As above, but (*ii*) testing pre-selected p and q. Use (6.4) to construct a table of probabilities (such as Table 6.2). The nth row of Pascal's Triangle can be used to fill in the coefficients of the form ^nC_r, all you have to do is to compute the several $p^{n-r}q^r$ terms.

(*c*) As (*a*) but (*iii*) $n > 25$, either testing $p = q = \frac{1}{2}$, or pre-selected p and q so long as $npq > 9$. Compute z as in expression (6.7) above, and find the associated probability from Appendix Table A.

Matched pairs Essentially the same procedure as above. In the event of ties, reduce n by the number of tied scores, and continue as before.

Exercises

1. It is suggested that women are generally safer drivers than men, and the following item of evidence is adduced in favour of this proposition. Over a given period of time, it was observed that 600 men and 400 women drivers entered a motorway at a given point, travelling in a given direction. In the 15 km stretch before the next access point, fifty-one of these were involved in accidents, of whom forty-one were men and ten women. Discuss the hypothesis in the light of this datum. (The binomial test does not, of course, make use of all the data available; and the reader might care to attack this problem again, after reading Chapter 7, using the methods given therein.)

2. An interior decorator is uncertain whether to use colour scheme A or B for a particular layout. He makes two mock-ups, identical save for these colour schemes, and invites twenty-two persons to view them in his absence, so as not to influence their choice. Eleven see A then B, the others B then A. Each is asked to rate each mock-up on a five-point scale: 5 = very pleasant, 4 = pleasant, 3 = neutral, 2 = unpleasant, 1 = very unpleasant. The ratings of the twenty-two subjects are given in Table 6.E2. Discuss the choice in the light of these data.

Table 6.E2   Ratings of colour schemes A and B by twenty-two observers

Observer	A	B	Observer	A	B
1	3	3	12	2	2
2	1	1	13	4	2
3	4	3	14	5	3
4	1	1	15	4	3
5	5	3	16	4	2
6	2	1	17	4	3
7	5	4	18	5	2
8	3	2	19	5	2
9	5	4	20	2	3
10	3	3	21	3	3
11	3	2	22	1	3

3. A Serious Minded Cannibal claimed that, even when both were curried, human flesh was strongly preferable to that of goats. He prepared two fine cauldrons of high-powered curry, one of each, and invited his friends to express their preference. Nineteen out of twenty expressed a preference for the human curry. Discuss this result.

4. A chess enthusiast asserts strongly that there is nothing to choose on current form between Spassky and Fischer. It is pointed out that in the

1972 championship match F won seven games, S won two, eleven were drawn, and one was awarded to S by default. He declines to change his mind. Is he justified? Are you justified in using the binomial test on these data?

References

Hammerton, M. "Measures for the efficiency of simulators as training devices", *Ergonomics*, 1967, **10**, 63–5.

Poulton, E. C. and Freeman, P. "Asymmetrical transfer effects in balanced experimental designs", *Psychol. Bulletin*, 1966, **66**, 1–8.

7

The uses of χ^2 - and others

We now come to discuss one of the most widely used of all the statistical tests available to the behavioural scientist. This is the χ^2 test. χ is the Greek letter chi, transliterated into English as CH — as in aCHilles. (It is pronounced to rhyme with "my", the ch rendered as in the Scottish "loch" — or simply as a K if you can't manage that.)

When I describe it as widely used this is no exaggeration: it is rare indeed to find a journal of the behavioural sciences which does not include at least one instance of its use. Nevertheless it has been denounced as unsound (Bradley, 1968) or at least as frequently misused (Lewis and Burke, 1949). The controversy about these contentions is important but somewhat rarified; and I will only say that it appears to me that the test is acceptable if certain precautions are observed.

When discussing the binomial test, we considered cases wherein observations were supposed to fall into one of two mutually exclusive classes. Suppose, however, that — as often happens — the number of such classes is not two but k $(k > 2)$. We may have some theory which predicts the relative frequency of such classes. Specifically, let us suppose that we predict that the probability of an observation being in the ith class

$(i = i \ldots k)$ is p_i; where $\sum\limits_{i=1}^{k} p_i = 1$. If the total number of observations

made is n, then the most probable value of the number of observations in the ith class is np_i. This is the *Expected Frequency* of the ith category,

which we will call E_i; and clearly $\sum\limits_{i=1}^{k} E_i = n$.

Now it is not often that we observe exactly what we expect; indeed, we may become highly suspicious if we do! Let us suppose that the actual frequency observed in the ith category is O_i. There will, in general, be some deviations from the several E_i of the O_i; are they enough to make us reject our hypothesis that the probability of an observation in the ith class is p_i?

In any category, the difference between the observed and expected values is evidently $(O_i - E_i)$ and since this can be negative it might be thought worthwhile to work with the square of the difference $(O_i - E_i)^2$ in each case. Now evidently an $(O_i - E_i)^2$ of (say) 10 is more impressive when E_i is (say) 3 than when it is (say) 100. This suggests that the quantity $\dfrac{(O_i - E_i)^2}{E_i}$ might be a reasonable quantity to use when examining our working hypothesis. The pioneer statistician K. Pearson suggested, in 1900, as an overall measure for the extent to which the observations differed from expectation the expression:

$$\chi^2 = \sum_{i=1}^{k} \frac{(O_i - E_i)^2}{E_i} \tag{7.1}$$

Evidently, the larger this is, the less well do the several O_i match the E_i. Further, however, Pearson showed that if the working hypothesis is true, χ^2 followed a certain well-studied distribution pattern known as chi-square. (To reduce the sufficiently ample confusion, the test we use is written as χ^2 while the name of the theoretical distribution is written out as "chi-square".) Tables can be constructed — as Appendix Table D in this book — which give the probabilities of χ^2 as large as those found, subject to certain conditions.

Before we examine some specific cases and their pitfalls, there is an important point to consider: this is the number of "degrees of freedom" (written d.f.) of a given χ^2. We have supposed there are k categories containing altogether n observations. Supposing we are told how many there are in the first category, we are still largely uncertain about how many there are in each of the others. In general, indeed, some uncertainty remains until we have been told how many there are in $k - 1$ categories. Then, however, no uncertainty remains; for since we know that the grand total is n, all those unaccounted for must fall into the final category: we have no "freedom" left to assign numbers. It is said, therefore, that in a simple k-category situation of this kind there are $(k - 1)$ d.f. This is important, for the quantity defined by chi-square, and hence the test value in χ^2, depends strongly upon the number of d.f.

It is time to clarify our ideas by considering an example.

Let us consider a very simple case of throwing an ordinary six-sided die. It is thrown, let us say, 240 times, so that our expected value for each of the outcomes 1–6 is 40: in other words, each E_i, $i = 1 \ldots 6$, = 40. Naturally, we do not observe exactly 40 for each outcome in practice, and our actual observations are set out in Table 7.1.

Table 7.1   Results of throwing a die 240 times

Outcome	1	2	3	4	5	6
O_i	42	50	38	30	45	35
E_i	40	40	40	40	40	40
$O_i - E_i$	2	10	2	10	5	5

We wish to know: are these data consistent with the hypothesis that the die is fair? The results do not look too bad, so we might not be called un-reasonable if we chose a somewhat slack α of 0.10 (so long as we are not going to play for high stakes).

Substituting in 7.1 gives:

$$\chi^2 = \frac{4}{40} + \frac{100}{40} + \frac{4}{40} + \frac{100}{40} + \frac{25}{40} + \frac{25}{40}$$

$$= \frac{258}{40}$$

$$= 6.45$$

and, from the argument above, there must be 5 d.f.

Consulting Appendix Table D, we find that the probability of a χ^2 as great as 6.45, given H_0, lies between 0.2 and 0.3. In fact, we were not, it seems, deceiving ourselves in thinking the die looked all right: H_0 is not to be rejected on our α, and the game may commence.

Dice are all very well; but more interesting problems can arise from field studies.

One of the leading pioneers of statistical thought was a splendid Great Victorian (capital letters essential) named Francis Galton. One of the important questions to which he bent his great intellect was: does the proportion of beautiful girls in the population vary from place to place within the country? Let us consider a modified version of the field study which Galton actually carried out. The experimenter travels successively to (let us say) seven cities: London, Manchester, Newcastle, Hull, Sheffield, Coventry and Oxford. In each one, at the same hour of the same day of the week he selects a vantage point where he sits with a counter in each hand. Whenever a girl passes he presses the counter in his left hand; and if he regards her as beautiful (on an admittedly subjective criterion) he presses that in his right hand also. When the total number registered — i.e., that on the left-hand counter — is 200 (say) he stops and goes home. He has thus obtained (he hopes) a random sample of 200 girls from each city, of whom some other lesser number are regarded as beautiful. The question is: does the proportion vary from city to city? The results obtained might look somewhat thus:

Table 7.2 Beauty frequencies in seven cities

City	L	M	N	H	S	C	O
No. of beautiful girls in first 200 sighted	15	8	8	6	10	6	17
No. of non-beautiful girls in first 200 sighted	185	192	192	194	190	194	183

Now obviously enough our H_0 is that the frequencies are in fact the same over the seven cities: in other words, we expect from each sample about the same number of beautiful and non-beautiful girls (since the total number in each sample is the same). Since there are altogether 70 beautiful and 1330 non-beautiful girls observed, our expectation must be that each city would provide 10 of the former and 190 of the latter. If we denote the number of beautiful girls observed in the ith city by b_i, and the number of the others by u_i, then substitution in 7.1 gives:

$$\chi^2 = \sum_{i=1}^{7} \frac{(b_i - 10)^2}{10} + \sum_{i=1}^{7} \frac{(u_i - 190)^2}{190} = 11.4 + 0.6 = 12.0$$

There are 6 degrees of freedom. (We can see that this must be so if we consider the following argument. In each column there are a total of 200 readings; therefore, as soon as b_i is known, there is no freedom left for n_i. Similarly, given Σb_i, there are 6 d.f. among the b_i, and hence 6 d.f. altogether.)

Consulting Appendix Table D, we find that the probability (two-tailed) of a χ^2 as large as 12.0 lies between 0.10 and 0.05; so whether we reject the *a priori* expectation of equal frequency must now depend on the α we thought reasonable. If (as seems fair enough to me, though the reader may differ) we require that our chance of being right should be as low as 0.01 before we agree that we are wrong, this result would not shift us. On the other hand, had we selected an α of 0.1, we would be persuaded — as Galton was, by the way — that there are indeed systematic differences in the proportion of beautiful girls in different places.

An important point emerges if we pursue our imaginary examples a little further. Our imaginary observer we allowed to be rather easily pleased, and he allotted no less than 5 per cent of his total sample to the category "beautiful". Suppose, however, that he had adopted a somewhat sterner criterion so that only fourteen girls altogether had been so categorized, and that none at all had been observed in H or C, two in M, and three in each of the others. Commonsense would suggest that, when expectations were so low, and the total likewise, not much significance could be attached to the differences. Commonsense, in this case, would be entirely right.

Very serious theoretical difficulties arise in handling χ^2 when the E-values are low; and there is much dispute over the reliability of the various practical minima which have been suggested. A good rule of thumb is: do not use χ^2 if any $E_i < 5$. However, if the number of d.f. > 5, it is acceptable to use the test if not more than one E_i in every 5 is <5 and if none of them is <1.

As a general rule, if these criteria are violated, χ^2 should not be used; but there may be a way round. We have so far supposed that we are always dealing with truly independent categories; but there are occasions when some of them can be combined, and the expectations therefrom pooled. This, however, is not a procedure to be undertaken lightly, or without careful thought.

Very roughly, we may say that categories may be pooled only when the resulting categorization is one which could well have occurred naturally in the first place. It may not be done if the result is a merely arbitrary aggregation of categories, which do not reasonably belong together.

This can be illustrated by a couple of imaginary examples. Suppose that there is a running track with six lanes set around the outside of a large oval. We wish to test the expectation that lane position is irrelevant to final position, i.e., that the probability of a runner coming first is the same whichever lane he is in. We find, however, that we only have results of twenty-four races, so that the expected number of winners in each track (if the null hypothesis is true) is always four. It seems to me that it would be quite reasonable to group the lanes into three sets of two: inner pair, middle pair, and outer pair, giving three expectations of eight each; for it would be quite plausible so to divide the set in the first place.

Suppose, on the other hand, we were investigating a prediction concerning the proportion of sick-leave at a certain factory over a given period of time, associated with various illnesses. Let us suppose that our E values for dental extractions, childbirth and sprained ankles were only 4, 3 and 2 respectively. There could be no earthly justification for pooling these to give a single category with an E of 9, since the groups simply do not hang together.

When used to examine a specific hypothesis, χ^2 is said to be used as a test of "Goodness of Fit". This term, surprisingly, means what it says: we are testing the "goodness" of the fit of the data to the theory. An example of this use can be obtained retrospectively, by seeing how it would have confirmed one of the classic experiments in plant genetics. The long-neglected pioneer of genetics, Gregor Mendel, bred some crossed varieties of peas, and as a result of his theories predicted the proportions of certain features in the next crop. The features he examined (the categories observed) were smooth-and-yellow, crinkly-and-yellow, smooth-and-green and crinkly-and-green. The probabilities (p_i) his theory gave for these were $\frac{9}{16}$, $\frac{3}{16}$, $\frac{3}{16}$ and $\frac{1}{16}$ respectively. The test crop consisted of $n = 556$ peas, and the observed and expected values (the latter rounded to the nearest unit) are given in Table 7.3.

Table 7.3   Results of Mendel's experiment on cross-bred peas

Category	Smooth/ yellow	Crinkly/ yellow	Smooth/ green	Crinkly/ green
$E_i = np_i$	313	104	104	35
O_i	315	101	108	32

Even at first glance, this is a truly beautiful fit. (There has, indeed, been a lot of discussion about how Mendel was fortunate enough to get results as good as these.) We can easily do the computation:

$$\chi^2 = \frac{4}{313} + \frac{9}{104} + \frac{16}{104} + \frac{9}{35} = 0.51$$

Evidently there are 3 d.f., and hence Appendix Table D shows that the probability of a χ^2 as large as this on Mendel's theory is between 0.90 and 0.95. The theory is well confirmed.

Quite generally, whenever data fall into frequencies in definable categories, χ^2 may be used to see whether there is any noteworthy difference between independent groups. In general, if r independent groups fall into k independent categories, we may inquire whether the frequencies with which the several groups are represented in the categories are different. Effectively, what we do is to test the hypothesis (called the "null" hypothesis) that all differences are due to random fluctuations.

Let us consider another imaginary case. An experimenter has read that amusing work "Gentlemen prefer Blondes", and decides to see whether the colour of a girl's hair is indeed determinant of her general attractiveness. He has therefore gathered groups of girls describable as blonde (B), brunette (D) or red-haired (R). These have been classified by a team of observers into the categories attractive (+), neutral (0) or repulsive (−).

The null hypothesis under test is, we may suppose, that the proportion of girls classified as (+), (0) and (−) are the same in the three groups. Table 7.4 shows the frequencies with which the three groups were

Table 7.4   Categorization of three groups of subjects

Group	B	D	R	Totals
Classification: +	11	12	10	33
0	19	20	15	54
−	20	13	10	43
Totals	50	45	35	130

represented in the three categories. The reader will note that, besides the 3 x 3 cells of the frequency table, we have noted the row and column totals, and the overall total, which is clearly both the total number of girls in the three groups and the total number in the three categories.

In order to use χ^2 we have to have expected (*E*) values for the several cells: how are these obtained?

Let us consider the top left-hand cell, the number of + blondes. Clearly, if the null hypothesis is true, we expect the proportion of members of the B column who are in the + row to be the same as the overall proportion of +s in the population. Evidently this *proportion* is equal to the total number of +s (i.e., the row total 33), divided by the size of the population (i.e., by the overall total 130). And this proportion is, by the same argument, the proportion of members of the B column we expect to find as +; so the expected value is the proportion multiplied by the column total (50). So the *E* value is:

$$E_{1,1} = \frac{\text{Row total} \times \text{Column total}}{\text{Population total}} = \frac{33 \cdot 50}{130} = 12.7$$

Generally, the expected value of the cell formed by the *i*th row of the *j*th column of an array of *r* rows and *k* columns is:

$$E_{ij} = \frac{\sum_{i=1}^{r} O_i \sum_{j=1}^{k} O_j}{\sum_{i=1}^{r} \sum_{j=1}^{k} O_{ij}} \tag{7.2}$$

i.e., it is the *product of the relevant row and column totals divided by the total number of items in the population*. We can now fill in the several *E*

Table 7.5 Expected values for contingency Table 7.4

Group	B	D	R
Classification: +	12.7	11.4	8.9
0	20.8	18.7	14.5
−	16.5	14.9	11.6

values in Table 7.5. We can now compute χ^2, by the most obvious extension of the expression given above. It is:

$$\chi^2 = \sum_{i=1}^{r} \sum_{j=1}^{k} \frac{(O_{ij} - E_{ij})^2}{E_{ij}} \qquad (7.3)$$

The two Σs merely indicate that all rows and columns are to be summed over — i.e., every cell is used. In our imaginary example this gives us:

$$\chi^2 = \frac{(11-12.7)^2}{12.7} + \frac{(12-11.4)^2}{11.4} + \frac{(10-8.9)^2}{8.9} + \frac{(19-20.8)^2}{20.8}$$

$$+ \frac{(20-18.7)^2}{18.7} + \frac{(15-14.5)^2}{14.5} + \frac{(20-16.5)^2}{16.5} + \frac{(13-14.9)^2}{14.9}$$

$$+ \frac{(10-11.6)^2}{11.6}$$

$$\doteq 2.13$$

There remains, of course, the question of how many d.f. are associated with this value. Fortunately, this is not too difficult. If there are r rows and k columns in a table, then for a given overall total there must be $(rk - 1)$ parameters to specify whatever general hypothesis is under discussion. From this, however, we must subtract the number of independent parameters needed to specify the null hypothesis, which (since H_0 requires that rows and columns are independent) is $[(r - 1) + (k - 1)]$. Thus the total number of d.f. is:

$$(rk - 1) - [(r - 1) + (k - 1)] = rk - r - k + 1 = (r - 1)(k - 1)$$

An argument which leads to this result, and which — though displeasing to pedants — is probably easier to understand, may be considered as follows.

In order to talk sensibly about the problem at all, we must know how many values fall in each row, and how many in each column. Thus, if we were filling in a particular row, we would have one fewer d.f. than there are cells in that row; and similarly, one fewer d.f. for a given column than there are cells in that column. Thus, for a table having r rows (categories) and k columns (groups) the number of d.f. is $(r - 1)(k - 1)$. In our case, this is $(3 - 1)(3 - 1) = 4$.

All that is left, then, is to look up Appendix Table D to find the probability of a chi-squared as large as, or larger than 2.13 on 4 d.f. We find that it is somewhat less than 0.7. In other words, this is by no means an unlikely result, and we have no call to reject our null hypothesis on any sensible α: there is no ground for supposing that gentlemen prefer blondes.

A case which occurs quite often in practice is that where there are two groups, who are divided into two categories — i.e., a 2 x 2 contingency table. Expression (7.2) above can be applied, of course; but it is sometimes handy to use a single expression which bypasses the need to compute the various Es. Let the observed values in the table be A, B, C and D as shown in Table 7.6 and let $A + B + C + D = N$.

Table 7.6   A 2 x 2 contingency table

Groups	I	II
Categories: 1	A	C
2	B	C

Then in this special case:

$$\chi^2 = \frac{N\left(|AD - BC| - \dfrac{N}{2}\right)^2}{(A + B)(C + D)(A + C)(B + D)} \qquad (7.4)$$

(7.4) cannot be derived simply from (7.3) above, since it incorporates a correction for discontinuity. (This is known as Fisher's "exact" test.)

Obviously, in this case, there is always 1 d.f. The use of this expression is again readily demonstrated by a simple example.

Let us suppose that an experimenter has two groups of rats. Group I consists of piebald animals, Group II of albinos. He is inquiring whether these two groups show the same kind of apparent handedness preference that we imagined in Chapter 6. He runs them all through a T-maze, and notes the numbers in category 1 (turned left) and category 2 (turned right). The results are given in Table 7.7.

Table 7.7   Results of an imaginary experiment

Groups	I	II
Categories: 1	15	8
2	5	12

Let us suppose further that the experimenter is prepared to press on with this work so long as the null hypothesis has less than a 10 per cent chance of being correct, i.e., he sets an α of 0.1. Now (7.4) above gives:

$$\chi^2 = \frac{40(180 - 40 - 20)^2}{20 \cdot 20 \cdot 23 \cdot 17} = \frac{1440}{391} \doteq 3.7$$

Consulting Appendix Table D we find that the probability of so large a χ^2 lies between 0.1 and 0.05: α is satisfied, and work will continue, since there seems to be a real difference between Groups I and II.

It is often possible, and sometimes necessary, to supplement the information obtained from χ^2 or binomial tests. Let us suppose that either (or both) of these useful tests has failed to show anything noteworthy in our data. Does this mean that there is indeed nothing of interest? The reader will immediately realise that it does not: there *may* be nothing worth looking at; but perhaps we have been looking at the wrong things, or attempting to use these tests in cases where they do not properly apply. Let us consider an example where these tests are not applicable.

Suppose our imaginary rat-man had performed another experiment on two groups of twenty rats each, again looking at the direction in which each rat turned at the choice-point in his beloved T-maze. Suppose further that the turns made by successive individuals in the two groups were recorded, and appeared thus:

Group I: RRLRLLLRLLLRRLLLRRRL
(i.e., 9 R-turns and 11 L-turns)

Group II: RRRRRRRRRRLLLLLLLLLL
(i.e., 10 R-turns and 10 L-turns)

Now neither the binomial nor the χ^2 test would have indicated any difference between the two groups. But look, not at the *totals*, but at the sequences. Do you think that there is no difference? Neither do I. The sort of hypothesis which immediately comes to mind is that Group I consists of animals which behave more or less randomly, while the second group were responding to some stimulus (a smell? a light gradient?) which abruptly changed halfway through. How can we test this? One way is to use the RUNS test.

Let us write down the sequence given by Group I, inserting a space whenever an R is followed by an L and vice-versa. Thus:

RR   L   R   LLL   R   LLL   RR   LLL   RRR   L

We have split the sequence into successive *runs* of like events; and we note that, on an n of twenty individuals, the number r of *runs* of like events is ten. Obviously, for Group II, $n = 20$ and $r = 2$. How are we to compute the probability of these numbers of *runs* under a null hypothesis that changes in direction are purely random?

Surprisingly, this is quite easy in principle; though it can become somewhat tedious with large n and r. Evidently, the null hypothesis requires us to say that all the $n!$ possible sequences of n observations are equally likely. In general, let the n observations be categorized as either xs or ys, and let us further suppose that there are m_1 xs and m_2 ys ($m_1 + m_2 = n$, obviously). From Chapter 3 it is clear that the number of possible arrangements of m_1 xs and m_2 ys is $^nC_{m_1}$. Let us now suppose that there are r_x runs of xs, and r_y runs of ys. Evidently, since the whole series can begin and end with either an x or a y, and, since every run of xs is terminated by a y, and vice versa, there are three possibilities to consider:

(1) $r_x = r_y + 1$

(2) $r_x = r_y$

(3) $r_x = r_y - 1$.

Case (1), of course, is that where the whole sequence of n observations both begins and ends with an x. The xs, we have decided, arrive in r separate groups, so ys must be inserted in $r_x - 1$ places to be found in the $m_1 - 1$ possible gaps between the xs. This, of course, can be done in $^{m_1-1}C_{r_x-1}$ ways. Similarly, the ys can be distributed into those gaps in $^{m_2-1}C_{r_y-1}$ ways. Thus the particular number of runs found could be arrived at in

$$^{m_1-1}C_{r_x-1} \times {}^{m_2-1}C_{r_y-1} = M \text{ (say) different ways,}$$

and its probability, given the null hypothesis is

$$\frac{M}{^nC_{m_1}}$$

Cases (ii) and (iii) can be solved similarly, and the probability of various frequencies of runs tabulated − straightforwardly if tediously.

Clearly, deviations from the values expected from the null hypothesis can occur in two ways: there can be *more* runs than are likely, and there can be *less*. The results of computing these boundaries are given in Appendix Tables E_1 and E_2. In the former, if we have m_1 observations in one category, and m_2 in the other, the entry against each pair is such that any number of runs equal to or less than r has a probability of <0.05 on the null hypothesis. In Appendix Table E_2, any number of runs $\geqslant r$ has a probability <0.05 on that hypothesis. (This is one of those cases where one's choice of α is effectively fixed by the published tables.)

We can now have another look at the result of our imaginary rat data. In the first group, $m_1 = 11$, $m_2 = 9$, $r = 10$. We quickly see from the tables that the limits for an α of 0.05 are $5 < r < 16$ so this result is unexceptionable. The second group, however, has an $r \leqslant 5$, and the probability of two or fewer runs is small on the null hypothesis: there is something worth looking at here.

The reader will have noted that the tables for this test are necessarily limited in extent. What is to be done when the numbers found in experiments lie outside these ranges? Fortunately, as this happens, the distribution of the runs probability approaches ever more closely to our old standby the Gaussian distribution. For large m_1 and m_2, the distribution approaches a Gaussian with a mean

$$\mu_r = \frac{2m_1 m_2}{m_1 + m_2} + 1 \tag{7.5}$$

and s.d.

$$\sigma_r = \sqrt{\left[\frac{2m_1 m_2(2m_1 m_2 - m_1 - m_2)}{(m_1 + m_2)^2(m_1 + m_2 - 1)}\right]} \tag{7.6}$$

so, to use Appendix Table A, we transform our results from r, m_1 and m_2 to the new variable z given by

$$z = \frac{r - \mu_r}{\sigma_r} \tag{7.7}$$

which involves a fair amount of arithmetic, but little real trouble.

In this chapter we have seen how to use the χ^2 statistic to examine the probability that the distribution of observations from *r groups* into *k categories* varies only randomly from group to group. It can also be used to examine the closeness with which such empirical results fit a theory, in which case it is being used as a test of goodness of fit. In all cases, the expected values (E) in each cell must be found, as well as the observed values (O). When used as a test of goodness of fit, the theory under consideration will, of course, give the several E_{ij}; in its more general use they can be computed as the product of the relevant row and column totals divided by the total number of observations. χ^2 can then be computed using (7.2) above.

The distribution of χ^2 depends upon the number of degrees of freedom in the situation; in the cases we have been considering this number is $(r-1)(k-1)$.

There are limiting cases where computation can be simplified. Specifically, where $r = k = 2$, (7.4) can be used — always with 1 d.f.

There is an important limitation on the use of χ^2. If the number of cells $\leqslant 5$, it should not be used if any $E_i < 5$; if the number of cells > 5, it should not be used if any $E_i < 1$, or if more than 20 per cent of Es are $\leqslant 5$. Commonsense and caution must be used in pooling categories.

It may sometimes happen that sequences of observations do not differ from one another in the frequency with which categories appear, but do differ in the way in which they are grouped.

Similarly, we may suspect non-random grouping in a single sequence of observations which fall into two categories. The randomness of a sequence of $m_1 + m_2$ observations, grouped into r runs of like sorts may be tested using the runs test, which is tabulated for fixed α.

Exercises

1. The diary of Dr Samuel Letsome, physician, contains the following entry for 31 February 1752: "In the morning to my Lord Fitzbooby's, to adminifter Phyfick. Found his Lordfhip far Gone with *Drooling Atrophy* of ye *Brain*, he confuming Brandy in vaft Beakers, tho' at his Laft Gafp. His brother, the Hon. Mr. Edward F., Likewife in dire *Straits*, swilling difh after difh of Tay, in great Agony with ye *Feftering Gut-Rot* . . . Pondered whether the *Favourite Tipple* of my Patients be related to their Difeafes. Made a *Table* . . . ." The table, somewhat simplified, appears thus:

| | Favourite drink | | | |
	Coffee	Rum	Brandy	Tea
No. of sufferers from DAB	6	22	7	0
No. of sufferers from FGR	35	3	2	9

What light can the reader cast on the learned doctor's speculations?
2. There were a number of occasions during the Second World War when the Luftwaffe assembled over 2000 aircraft for a specific operation. The distribution by function is given in the following table.

Operation	Yellow May '40	Eagle July '40	Barbarossa June '41	Rhine-Watch Dec. '44
Medium bombers	1330	1330	780	100
Dive/Strike	380	280	310	390
Single-engine fighters	860	760	830	1770
Twin-engine fighters	350	250	90	140
Reconnaissance	640	170	710	60

Examine and comment on the changes in distribution.

3. Passing a queue at the University Graduate Centre recently, I noticed the sequence of men (M) and women (W), which was as follows:

MWMWMMWMWMWMWMWWMWMMWMWMWMWMM

Comment.

4. During the French Revolutionary and Napoleonic Wars, the Royal Navy achieved four major victories in Fleet actions in the open sea. The following table gives the results:

	Battle			
	Ushant	*St Vincent*	*Camperdown*	*Trafalgar*
Commander	Howe	Jervis	Duncan	Nelson
No. of enemy ships engaged	26	27	16	33
No. taken or sunk	7	4	8	20

Do these data suggest that the commanders were not equally successful?

References

Bradley, J. V. *Distribution free statistical tests*, Prentice-Hall, 1968.

Lewis, D. and Burke, C. J. "The use and misuse of the chi-square test", *Psych. Bull.*, 1949, 46, 433—98.

Pearson, K. "On the criteria that a given system of deviations . . . (can be) supposed to have arisen from random sampling", *Phil. Mag. Series 5*, 1900, **50**, 157—72.

8

Comparing ranks

When discussing the binomial and χ^2 tests, we have been concerned essentially with data which can be merely classified. We have considered obvious dichotomies, such as dead/alive, male/female; and we have discussed categorization into groups such as attractive, neutral, repulsive. These are still very broad groupings; and we have not asked what can be done with information — which we often have — about relative position within groups, or according to some variable character.

It is usually the case that when extra information is available it can be used; and so it is here. In this chapter we shall discuss some of the information which can be extracted when two independent samples can be ranked for some characteristic.

The tests which we will be discussing in this chapter are based on the method of randomization, which was first adumbrated by R. A. Fisher. In principle, this method is simple enough; and although its application to specific problems often involves some formidable algebra, the tests which are then derived from it are distinguished by great simplicity, handiness and power. In fact, we have already considered a particular case of the method, in the runs test; and the principle of the method is always the same: assuming the null hypothesis, *all possible* outcomes are considered, and the probability p of the *actual* outcome (or of any equally or less likely) is computed. The null hypothesis H_0 is regarded as rejected if the value of p is less than that of our selected α.

As usual, this concept becomes clearer when we consider a specific application. Let us suppose that we measure the performance of two closely comparable groups of subjects on some task: one group, numbering n, in loud noise, the other, numbering m, in quiet conditions. Let us further take all the $(n + m)$ scores so obtained, and give to the smallest score obtained the rank 1, to the next smallest the rank 2, and so on to the largest which will have rank $n + m$.

Let us now arrange the scores in this rank order, labelling each rank N

if it was obtained in noise, and Q if in quiet. We might get a sequence somewhat thus:

$$1(N), 2(N), 3(Q), 4(N), \ldots r(N), r + 1(Q) \ldots n + m(Q)$$

It will at once be evident to the reader that we have a situation analogous to that which lead us to the runs test; but now we are asking a different question, namely: do the Ns and Qs tend to cluster towards opposite ends of the sequence? Or, in other words; do the conditions of noise and quiet make a substantial difference?

What statistic shall we use? One which seems worth trying is R_n, i.e., the total of all the *n ranks* obtained in condition N (or alternatively R_m, the total of the m ranks obtained in condition Q). This does turn out to be useful, when used in conjunction with the related quantity U (studied by Wilcoxon and by Mann and Whitney). U is defined as the number of Ns which are preceded by all Qs when they are set out as above. Thus, if we consider the sequence $QNQNQNN$ the first Q has no preceding Ns, the second has 1 and the third 2 (i.e., those with ranks 2 and 4). Therefore the value of U for this sequence is $0 + 1 + 2 = 3$.

It is obvious that more than one sequence of the same number of Qs and Ns can give rise to the same value of U. For example, the two series $NNQQNN$ and $NQNNQN$ both yield a U of 4. In general, if we have a sequence of n items of one sort (xs, let us say) and M of another (ys) let $S_{n,m}(U)$ be the number of arrangements wherein ys preceed xs exactly U times. Now evidently the total number of places in the sequence is $(n + m)$ of which m are occupied by ys; therefore the total number of possible sequences is equal to the number of ways of selecting m places out of $(n + m)$; i.e., the total number of possible sequences is $[(n + m)!]/[n! \, m!]$. Given that all sequences are equally likely — which is equivalent to the null hypothesis H_0 — we can now state the probability of a given U. It is:

$$p(U) = \frac{S_{nm}(U)n! \, m!}{(n + m)!} \tag{8.1}$$

This quantity has been computed for a range of values of n and m. Let us examine an application of this test.

On the day on which I am writing this, I completed the first part of an experiment in which I am examining the effect of noise on a discrimination task. My experimental group (E) performed the task once, and on a later day repeated it in noise, my control group (C) repeated it in the same quiet condition as they had originally. There is, of course, a practice effect to be expected — people tend to be better the second time — but my prediction was that the Es would do relatively, and perhaps absolutely worse, the second time. The reader will note that this is a one-tailed prediction; and, being a trifle sceptical, I set a moderately strict α of 0.025. Table 8.1 gives the difference between each Ss score on his first and on his second trial: a positive value is an improvement; a negative one a decrement.

Table 8.1   Results of an experiment on the effects of noise

Subjects		Difference score	Rank of score
Control	1	0.14	10
	2	−0.04	5
	3	0.07	8
	4	0.08	9
Experimental	1	0.04	7
	2	−0.12	2
	3	−0.08	4
	4	0.02	6
	5	−0.11	3
	6	−0.14	1

The scores are ranked in order of size, smallest first. The smallest score (here actually negative) is obviously −0.14; and we can write down the rank sequence thus obtained. Clearly, it is:

EEEECEECCC

Since our hypothesis is that the Es are smaller, we determine U for the Es. Since only those with ranks 6 and 7 are preceded by Cs — one in each case — the value of U is 2 on an n of 4 and an m of 6. Appendix Table F gives the one-tailed probability of a U as small as 2 as 0.019. (For a two-tailed question, double the p-values in the table.) Our α is satisfied; it really seems that there may be something in this; so I shall proceed with part two of the experiment.

No doubt the reader will have already thought that the process of counting predecessors might become a bit tedious, and possibly somewhat prone to error, with n or m values larger than 6 or 7. Fortunately there is a method of avoiding this. We remarked above that the quantity U was associated with the rank total of the two groups. The relation is, in fact, the fairly simple one:

$$U = nm + \frac{n(n+1)}{2} - R_n \tag{8.2}$$

or.

$$U = nm + \frac{m(m+1)}{2} - R_m \qquad (8.3)$$

In general, of course, $8.2 \neq 8.3$, the larger of the quantities being not U but U'. If in doubt, it is always possible to ensure that we have the right one by using the expression:

$$U = nm - U' \qquad (8.4)$$

As usual, we can best clarify our thoughts by an example.

Some years ago I conducted an experiment in which subjects were trained upon various sorts of simulators to control a moving vehicle. I had reason to expect that simulator A would prove superior to simulator B; I had intended to use equal numbers in each trial group; but for various reasons ended up with 12 As and 11 Bs. For each subject, one of the scores obtained was a measure of the transfer of training from the simulator to the real thing; the larger the better. The actual results and their ranks are given in Table 8.2.

Table 8.2   Transfer scores (ranked) obtained by two experimental groups

Group A			Group B		
Subject	Transfer score	Rank	Subject	Transfer score	Rank
1	0.96	18	1	0.19	2
2	1.15	23	2	0.07	1
3	0.99	20	3	0.22	3
4	1.08	22	4	0.40	8
5	0.65	11	5	0.62	10
6	0.40	8	6	0.89	16
7	0.74	12	7	0.40	8
8	1.03	21	8	0.32	6
9	0.98	19	9	0.92	17
10	0.88	15	10	0.31	5
11	0.81	14	11	0.25	4
12	0.76	13	—	—	—
	Sum of ranks:	196		Sum of ranks:	80

The first thing the reader will note is that three subjects achieved the same score, that of 0.40. What should be done about ties? We shall return to this point later; for the time being it will suffice to say that each score is given the mean of the ranks which would have been distributed had the tied scores been discriminable. Thus, here there would have been ranks of 7, 8 and 9 to distribute, so all are awarded rank 8.

As stated above, I was asking a one-tailed question; and I was fairly confident that my opinions were sound. I was quite prepared to accept an α of 0.05. Substituting in (8.2) above, we have (since $n = 12$, $m = 11$, and $R = 196$):

$$U = 12 \cdot 11 + \frac{12(12 + 1)}{2} - 196$$

$$= 132 + 78 - 196$$

$$= 14$$

Here (in case we are nervous) we apply (8.4) and find that $nm - U = 118$, so (since we always want the *smaller* value) our U of 14 is correct. Consulting the tables we find that $p \mid H_0) < 0.025$, so we feel pleased with ourselves and get on with the job. (Strictly, the significance levels given in tables are only approximate if there are tied scores. This only matters if the ties are very numerous, and clearly does not here.)

The tables only take us as far as $n = m = 20$; what are we to do with larger groups? This question was studied by Mann and Whitney (1947) who showed that, as n and m increased, the distribution of U approached ever more closely to a Gaussian distribution with mean

$$\mu = \frac{nm}{2}$$

and s.d.

$$\sigma = \sqrt{\left[\frac{nm(n + m + 1)}{12} \right]} \, .$$

Thus, we can use our table of the Unit-Normal (Appendix Table A) for larger n and m by transforming to a new variable z given by:

$$z = \frac{U - \dfrac{nm}{2}}{\sqrt{\left[\dfrac{nm(n + m + 1)}{12} \right]}} \tag{8.5}$$

U being computed using (8.2) as usual. The arithmetic can become a little tiresome here, but never really overwhelming. Let us, without filling in any background, simply imagine a case. Let $n = 16$, $R_n = 210$, $m = 24$, $R_m = 610$.

We substitute in (8.2), giving

$$U = 16 \cdot 24 + \frac{16 \cdot 17}{2} - 210 = 310$$

At this point (if we have any sense at all) we feel a chill doubt, check using (8.4) and conclude that we have arrived at U' by mistake: in fact

$$U = 16 \cdot 24 - U' = 384 - 310 = 74$$

We can now substitute in (8.5), finding that:

$$z = \frac{74 - 192}{\sqrt{(32.41)}} = \frac{118}{\sqrt{(1312)}} = 3.26$$

Appendix Table A gives the probability of a z as extreme as, or more extreme than this as <0.001; which should satisfy all but the most rigorous choices of α.

Above, we discussed corrections for tied ranks. Strictly speaking, the distribution of the mathematical model underlying the Gaussian transformation of the U-test implies that all ties are due to imperfections of measurement; but an adjustment to (8.5) has been developed to allow for their appearance. Suppose that there are t ties for a given rank, then the quantity $T = (t^3 - t)/12$ is formed. There may be more than one rank with ties, each of which will have its associated T; then let ΣT be the sum of all such terms. Let $K = m + n$, then the modified form of (8.5) is

$$z = \frac{U - \dfrac{nm}{2}}{\sqrt{\left[\dfrac{nm}{K(K-1)} \left(\dfrac{K^3 - K}{12} - \Sigma T \right) \right]}} \tag{8.6}$$

There is, however, a curiosity associated with the use of (8.6): it tends to increase the value of z; i.e., it makes it easier to satisfy a given α. Perhaps justly cautious workers might hesitate to use (8.6) rather than (8.5); but it is to be recommended if there are large numbers of groups of ties, so long as the total number of readings is large enough to justify the use of the z transformation.

Let us examine another imaginary case. Two classes of schoolchildren have been taught English by the same teacher, and at the end of the year both classes are given the same examination. All the scripts are marked by an independent examiner; but, on glancing through the results, the headmaster entertains doubts as to whether equal success has been achieved by

Table 8.3   Results of an English examination

Class X			Class Y		
Child No.	Score	Rank	Child No.	Score	Rank
1	80	46	1	73	44
2	75	45	2	67	41
3	70	43	3	65	40
4	68	42	4	59	37½
5	62	39	5	56	36
6	59	37½	6	48	32
7	55	35	7	45	30
8	51	34	8	41	28
9	50	33	9	36	26
10	46	31	10	35	25
11	44	29	11	28	21
12	40	27	12	25	20
13	32	24	13	24	18
14	30	23	14	20	14
15	29	22	15	16	11
16	24	18	16	14	10
17	24	18	17	13	8
18	22	16	18	13	8
19	21	15	19	12	6
20	18	13	20	11	5
21	17	12	21	9	4
22	13	8	22	8	3
—	—	—	23	7	2
—	—	—	24	6	1
	Sum of ranks:	610½		Sum of ranks:	470½

the two classes. The exam scores are given in Table 8.3. It is at once evident that $n = 22$, $m = 24$ (hence $K = 46$), $R_n = 610.5$, $R_m = 470.5$. Whence (8.2) gives:

$$U = 22 \cdot 24 + \frac{22 \cdot 25}{2} - 610.5$$

$$= 192.5$$

However, we further note that there are numerous ties. Specifically there are three scores with rank 8, three with rank 18, and two with rank $37\frac{1}{2}$.

Thus we have t values of 3, 3 and 2; and thus the value of ΣT is:

$$\frac{3^3 - 3}{12} + \frac{3^3 - 3}{12} + \frac{2^3 - 2}{12} = \frac{24 + 24 + 6}{12} = 4.5$$

We may now substitute in (8.6) to find z. We have:

$$z = \frac{192.5 - \dfrac{22 \cdot 24}{2}}{\sqrt{\left[\dfrac{22 \cdot 24}{46 \cdot 45} \left(\dfrac{46^3 - 46}{12} - 4.5 \right) \right]}} \doteq 1.57$$

Whence, consulting Appendix Table A, we find that $P|H_0) = 0.06$.

Now whether the head regards this as sufficient grounds for deciding that there is indeed a difference between the classes would depend upon the value of α which he pre-set. What this was, we cannot say; for few and impious are those who seek to penetrate the minds (if any) of headmasters. The reader will readily see, also, that he would have a wide range of hypotheses to choose from, all reasonable enough *a priori*, in order to account for the difference.

We have now discussed at some length the slightly less straightforward forms for U when n or m is large, and when many ranks are tied. However, the reader should not let these aspects of the matter obscure the fact that the U test is one of the best, quickest and most reliable tests available. With a little practice, its use with small n or m can become very rapid indeed; and in the slightly longer case where n and m are in the range 9–20 it is still faster and handier than any comparable test.

In a later chapter we shall discuss the relative powers of tests; and at this point it can suffice to say that few are more powerful. Moreover its range of validity is very great. The only assumption concerning the underlying distributions of the data made is this: that the distributions of the two groups being compared may differ in location (that, after all, is what we are looking for), but are similar in form. That is to say, the distributions do not have to be described by any particular function, as long as both are described by the same, or at least very similar, functions. So long as the readings can be unambiguously set in order, the U test can then be employed. (Tests which have this characteristic of not demanding a specific form of distribution in the data are sometimes called "distribution free", and sometimes "non-parametric". For obvious reasons they are singularly well adapted to the human sciences.)

In an earlier chapter we mentioned the case when, in two batches of data, each item in one batch is "matched" with a corresponding item in the other. These "matched pairs" situations can arise, as we noted, when, e.g., each subject acts as his own "control" to the "experimental" situation. Although — it is worth reiterating — matched-pairs designs should be adopted with caution, there is an admirable distribution-free test adapted for them. This is the Wilcoxon T-test. Suppose that we wish to compare the effects upon some task of two conditions X and Y; and that to that end we have obtained n matched pairs of readings: one of each pair in each condition.

It is clear that, in this situation, the null hypothesis is that the two conditions do not have distinguishable effects on performance, and that differences between pairs of scores are merely random. Let us concentrate upon the differences between the two scores of each pair; let the ith pair have scores x_i in condition X and y_i in condition Y. If we now form the several $d_i = (x_i - y_i)$ we will have n terms, each of which will have some absolute magnitude $|d_i|$ and an algebraic sign s_i, which will be either $+$ or $-$, so long as $|d_i| \neq 0$. For example, if $x_i = 5$ and $y_i = 7$, d_i has absolute magnitude 2 and sign "$-$".

Let us now rank the magnitudes *neglecting signs*. We do this as for U, with the smallest magnitude having rank 1 and so on.

If H_0 is true, we would expect each magnitude $|d_i|$ of a given sign to be closely matched (indeed exactly matched, in a sufficiently large sample) by another magnitude of opposite sign. Thus, in a sample of finite n we would expect that if we added all the *ranks* which have the *same* sign, the sum would be about the same as the sum of all the ranks of the opposite sign. If we denote the sum of all ranks of one sign as $T = {}_{s_1}\Sigma R_i$ and that of the other by $T' = {}_{s_2}\Sigma R_i$ then H_0 predicts that $T \doteq T'$. It also suggests, of course, that the sign of any *given d* is as likely to be $+$ as $-$. Since, for n pairs, there are 2^n possible arrangements of $+s$ and $-s$ each of which will have some associated value of T, it is possible to compute the probability of any T for any n given H_0. This has been done.

Let us consider an example. A manufacturer of plastic kits wishes to know whether two forms of instructional leaflet, A and B, differ in clarity. Two kits, 1 and 2, closely matched for difficulty of assembly (or ease, if you prefer it) are prepared, and a group of ten volunteers is tested. Five of them are given Kit 1 with instruction A, followed by Kit 2 with B; the other half have these in reverse order. In all cases the time taken to assemble (in minutes) is recorded, this being the subject's score.

It is worth pausing a moment to note a couple of assumptions we have had to make. First, we assumed that the kits were equally easy to assemble; second, and rather more subtly, we assumed that reversing the order cancelled out learning effects. If either of these assumptions were unsound, the whole problem becomes much more complex; and such situations will be discussed in a later chapter.

The results of our imaginary experiment are shown in Table 8.4.

Table 8.4   Results of a comparison of two forms of instructions

Subject	Time taken (mins) Instruction A	Instruction B	$(B - A)$	Rank of $|(B - A)|$	Rank with least frequent sign
1	105	116	11	8	
2	100	126	26	10	
3	146	141	−5	4	4
4	59	66	7	6	
5	46	44	−2	1	1
6	84	87	3	2	
7	90	98	8	7	
8	77	73	−4	3	3
9	62	74	12	9	
10	75	81	6	5	

The importance to the manufacturers of the question we are asking is somewhat obscure; so let us arbitrarily fix α as 0.05. Obviously we are asking a two-tailed question; and we note that, of the 10 ds, only three have negative signs. When asking two-tailed questions, we naturally look for the less frequent sign; while a one-tailed question commits us in advance to the sign we are interested in. Here, the ranks associated with the less frequent sign are 1, 3 and 4. Hence $T = 1 + 3 + 4 = 8$. Consulting Appendix Table G we find that $p\,|\,H_0) < 0.05$ and we decide that there is something in it: instruction A really seems to lead to shorter times.

Whatever may be the pitfalls in setting up matched-pairs designs, the reader will, I imagine, readily concede that few tests rival T for sheer speed of application.

Sample sizes of $n > 25$ are not covered by our Table G; and if larger numbers are in contemplation a slightly more laborious arithmetic must be used.

The conclusions of the Central Limit Theorem will have led us to expect that the distribution of T might approach the Gaussian distribution as n becomes large. This is so, and the mean of this distribution is

$$\mu_T = \frac{n(n + 1)}{4} \text{ and its s.d. is}$$

$$\sigma_T = \sqrt{\left[\frac{n(n + 1)(2n + 1)}{24} \right]}$$

Hence for large n we compute T as before, and then transform to a new variable given by

$$z = \frac{T - \dfrac{n(n + 1)}{4}}{\sqrt{\left[\dfrac{n(n + 1)(2n + 1)}{24}\right]}} \tag{8.7}$$

and we can then use Appendix Table A for the associated probability under H_0. (It should be remarked that tied ranks are dealt with exactly as in U: each is allotted the mean position.)

Let us suppose a case where this method is useful. A researcher is doubtful whether two tests of reasoning power are indeed of equal difficulty. However, by a fortunate chance, he has an unusually good opportunity for examining the question. He has been working with a group of identical twins, all of whom are living normal lives, not separated at a tender age like so many twins in textbooks. Now, under these circumstances he would expect there to be little overall difference between the results if he gave one test to each twin, if H_0 were true; i.e., if the tests were of equal difficulty. He therefore gave the tests to thirty-two pairs of twins deciding by toss of coin which twin was to have which test. The results are given in Table 8.5.

Table 8.5   Results of a twin study

Pair No.	Score test 1	Score test 2	d (2−1)	Rank of |d|	Pair No.	Score test 1	Score test 2	d	Rank of |d|
1	31	38	7		17	81	92	12	
2	41	45	4	10½	18	96	94	−2	5*
3	59	62	3	8	19	83	98	15	
4	26	29	3	8	20	60	73	13	
5	53	61	8		21	95	93	−2	5*
6	58	61	3	8	22	7	44	37	
7	97	95	−2	5*	23	90	89	−1	2*
8	93	89	−4	10½*	24	37	38	1	2
9	59	65	6		25	35	60	25	
10	50	67	17		25	1	19	18	
11	29	37	8		27	9	29	20	
12	59	60	1	2	28	68	82	14	
13	53	62	9		29	32	45	13	
14	86	96	10		30	73	86	13	
15	94	89	−5	12*	31	29	50	21	
16	27	38	11		32	66	88	22	

It will immediately be noted that, of the thirty-two ranks, only six are negative so we need the sum of the ranks of negative sign. The reader will also note that the columns of ranks of $|d|$ have only been partially filled in. This is mere commonsense: once we have obtained all the ranks we need — those of negative sign — there is no point whatever in going to the labour of finding the rest. (The required ranks have been marked *.)

We easily find that $T = 2 + 5 + 5 + 12 + 10.5 + 5 = 39.5$.

We know that $n = 32$, whence, substituting in (8.7):

$$z = \frac{39.5 - 8.33}{\sqrt{\left(\frac{32 \cdot 33 \cdot 65}{24}\right)}} = \frac{-224.5}{2\sqrt{(715)}} \doteq 4.2$$

Appendix Table A indicates that this has a p-value under H_0 of <0.001. We did not decide what α our researcher would decide upon; but it would be a cautious man indeed who would not decide that there was a genuine difference in the teeth of this result.

In this chapter we have discussed two tests: the U-test and the Wilcoxon T-test. Of all the weapons in the behavioural scientist's armoury, they are among the swiftest, the simplest, the most powerful and the most parsimonious. The reader would be well advised to practise them both until fully competent. Both answer the question: is there a difference between the median locations of these two groups? The U-test deals with independent samples; the T-test with matched pairs. Because of the difficulties and pitfalls associated with setting-up matched-pair situations, the U-test is the more widely usable of the two.

The use of U presents little difficulty. First, the scores of the two groups together are ranked for size, the smallest score being given rank 1. If the numbers in the two groups are small, U is determined simply by *counting predecessors* as we did in the first example discussed. If n and m (the numbers in the two groups) are in the range 9–20, U is determined by substitution in (8.2), always checking that the smaller of the two possible values has been formed (expression (8.4) provides the check). If either n or $m > 20$ (8.5) may be used to transform U to a Gaussian form; but should there be many tied scores, (8.6) should be used.

T is even simpler. Score the two sets of matched pairs, and find the *differences* between each pair of scores. Rank the *absolute magnitudes* of the differences, merely noting these signs. Add the ranks of *least frequent sign*. This is T. Should the number of pairs exceed twenty-five, T is to be computed in this way, and transformed to a Gaussian variable using (8.7).

Exercises

1. Two small groups of schoolchildren, matched by their teacher for
intelligence and ability, were set to learn a batch of names and dates. One
group (P) was told, by someone whom they believed, "If you don't do
well on the test to follow, I'll belt the bloody hide off you!" The other
(R) was told: "If you do well on the test to follow, you will get a free
ticket to a cinema." The actual scores are given in the table below.
Comment.

P-Group	92	86	85	78	62	61
R-Group	80	75	65	59	55	—

2. The judges of a cookery competition found it impossible to decide
between the two finalists X and Y. Each finalist was therefore required, on
two successive nights, to prepare the same five-course meal for five
persons. A group of ten gourmets was provided with these meals, half
having the meals in the order XY, the other in the order YX. Each course
was marked out of 20; so that each gourmet had eventually tasted both
meals and marked them out of 100. The results are given in the table
below. Comment.

	Gourmet									
	1	2	3	4	5	6	7	8	9	10
X-Score	66	23	51	68	92	14	62	59	78	50
Y-Score	65	23	48	53	99	15	60	45	70	40

3. In one part of an experiment now going on, my colleague Mr A. H.
Ticknert (who has kindly made these data available to me) is examining
the abilities of groups of subjects to detect incidents displayed on television screens under two conditions A and B. Two groups, each of thirty-
six subjects, have so far taken part in this experiment (I have given only
part of a very complex body of data), one group being assigned to each
condition. The table gives the number of detections made by the several
subjects. Comment.

† Mr Tickner died while this book was being prepared for Press.

Group 1 (Condition A)

Subject	1	2	3	4	5	6	7	8	9	10	11	12
Detections	4	4	2	5	5	6	1	6	3	5	3	4

S	13	14	15	16	17	18	19	20	21	22	23	24
Ds	2	4	3	6	3	3	1	3	4	4	4	3

S	25	26	27	28	29	30	31	32	33	34	35	36
Ds	3	4	4	5	5	4	5	4	4	3	5	0

Group 2 (Condition B)

Subject	1	2	3	4	5	6	7	8	9	10	11	12
Detections	3	2	6	4	4	3	4	3	5	6	6	4

S	13	14	15	16	17	18	19	20	21	22	23	24
Ds	4	3	7	4	7	5	3	4	6	2	3	4

S	25	26	27	28	29	30	31	32	33	34	35	36
Ds	1	7	6	6	4	3	5	3	5	5	6	1

4. A Fearfully Responsible Investigator hopes to demonstrate the dire effects of reading Depraved Works upon the Young. He finds twelve pairs of twins. These he contrives to separate one from the other; and he brings up one of each pair on an unrelieved diet of Improving Books, while the other he has lisping de Sade at a very early age. Many years later, both members of eleven pairs (one having escaped) were given a battery of tests devised by . . . (the reader may fill in some proper name here) to give a Randiness Index (R.I.). The results are given in the table.

	Twin pairs										
	1	2	3	4	5	6	7	8	9	10	11
R.I. of "pure" twin	79	79	46	19	21	52	77	7	48	45	66
R.I. of "depraved" twin	3	18	33	57	16	71	85	60	43	42	25

Has the F.R.I. proved his point?

References

Mann, H. and Whitney, D. "On a test of whether one of two random variables is stochastically larger than the other", *Ann. Math. Statist.*, 1947, **18**, 50—60.

Wilcoxon, F. *Some rapid approximate statistical procedures*. American Cyanamid Co., Stamford Conn., 1949.

9

More methods based on randomization

The U and Wilcoxon tests, powerful and handy though they are, have
certain inherent limitations. Essentially, they both compare two samples
and two samples only, to see if they derive from the same population.
Often, however, we would like to do more than this. It frequently happens
that we are presented, not with just two alternatives, but with a variety of
possibilities. As we have seen, if the measure we are considering divides the
population into broad categories we can use χ^2 in such a case; but when —
as often happens — scores fall upon an ordered continuum, we might
expect to be able to do better than that. Indeed, we can.

Suppose that half-a-dozen different positions of driver's seat and
steering-wheel are under consideration for a new long-distance lorry. We
might decide to test (say) eight subjects on each design, requiring each
subject to drive round a set course, and marking his performance on such
measures as time taken, distance by which obstacles were avoided, etc. We
would then have six sets, each of eight values of scores, and we wish to
know whether there are outstanding differences between them. Let us
further (to make things more realistic) suppose that two drivers failed, in
the event, to complete the course, so that we have four sets of eight, and
two of seven scores to compare.

The null hypothesis in any such case is evidently that all these forty-six
scores belong to the same population, and that there is only random
sampling variability among them.

Before attending further to this example, let us consider the general
case.

What we are concerned with is whether there is a difference between
K different treatments, one of which has been applied to each of K independent samples. Let us suppose that the first treatment has been applied
to a sample of size R_1, the ith treatment to a sample of size R_i; and so on.
Let us set out all the data in a table, each *column* being the scores obtained
with one particular treatment: we then have a table of K columns of

various lengths R_i. The total number of entries in the table will obviously
be $\sum\limits_{i=1}^{K} R_i = N$ (say).

Now let us rank all the N entries in the table from 1 to N in order of
increasing magnitude. Let the sum of the ranks in the ith column be T_i.
Then the mean of the ranks of the ith column must be $(T_i/R_i) = \overline{T}_i$.

Now the N ranks in the table constitute the first N integers; under H_0
all $N!$ possible tables are equally likely; and always the mean of the whole
table must be $(N + 1)/2$. Hence, also, if H_0 is true the *expected* value of T_i
must be $[(N+1)/2] R_i$ and of $\overline{T}_i$ simply $(N + 1)/2$ again.

Thus the deviation of a column mean from its expected value under
H_0 is

$$\overline{T}_i - \frac{N + 1}{2}$$

and its square is

$$\left[T_i - \frac{N + 1}{2} \right]^2$$

As soon as we stop and think, it must strike us that the importance of a
deviation is proportional to the size of the particular sample (remember
that they are not necessarily of equal size); and we might suggest using
such a statistic as

$$\sum_{i=1}^{K} R_i \left[\overline{T}_i - \frac{N + 1}{2} \right]^2$$

However, Kruskal and Wallis (1952) demonstrated the advantages of
using the somewhat more complex quantity H, defined by:

$$H = \frac{12}{N(N + 1)} \sum_{i=1}^{K} \left[T_i - \frac{(N + 1)^2}{2} \right] \tag{9.1}$$

which can be reduced to a computationally simpler form:

$$H = \left[\frac{12}{N(N + 1)} \sum_{i=1}^{K} \left(\frac{T_i^2}{R_i} \right) \right] - 3(N + 1) \tag{9.2}$$

Using the method of randomization, we can, in principle, compute the
probability of finding an H as large as a particular value. It can be computed by finding how many of the $N!$ possible tables give equal or larger
values, and dividing that number by $N!$. With the computers available
today, this is not very difficult; and Appendix Table H (appropriately,

though unintentionally) gives these values for some small Ns and Ks, and a few R_i. However (and herein lies the beauty of using the expression) Kruskal and Wallis were able to show that, as these quantities increase, H approaches ever more closely to χ^2 with $(K-1)$ degrees of freedom. Hence the χ^2 table (Appendix Table D) can be used in most practical cases.

We can now return to our lorries. Let us suppose that scores, and hence the ranks, obtained by the forty-six subjects distributed among the six groups were as given in Table 9.1.

Table 9.1 Scores (and ranks of scores) obtained from an investigation into cab design. (The rank of each score is given in brackets after the absolute value)

	Group					
	A	B	C	D	E	F
Subject 1	87 (38)	89 (39)	90 (40)	92 (41)	99 (46)	95 (43)
2	86 (37)	82 (36)	79 (35)	75 (34)	98 (45)	71 (32)
3	70 (31)	67 (29)	65 (27)	66 (28)	96 (44)	68 (30)
4	60 (26)	56 (24)	55 (23)	51 (22)	93 (42)	50 (21)
5	37 (16)	40 (17)	41 (18)	43 (19)	74 (33)	40 (20)
6	33 (15)	31 (13)	30 (12)	27 (11)	58 (25)	24 (10)
7	21 (8)	23 (9)	18 (6)	20 (7)	32 (14)	17 (5)
8	11 (1)	12 (2)	13 (3)	—	—	14 (4)
Rank sum T_i	172	169	164	162	249	165

Substituting in (9.2), we have

$$H = \frac{12}{46.47} \cdot \left[\frac{(172)^2}{8} + \frac{(169)^2}{8} + \frac{(164)^2}{8} + \frac{(162)^2}{7} + \frac{(249)^2}{7} + \frac{(165)^2}{8} \right] - 3.47$$

It must be admitted that the arithmetic is a trifle laborious if you have only pencil and paper; but this falls out to: $H \doteq 239 - 141 = 98$. Since $k = 6$, we consult Appendix Table D for $\chi^2 = 98$ with 5 (i.e., $6 - 1$) d.f. We find that $p \mid H_0) < 0.001$. We did not assign an α value in advance; for without knowing how important the result was to the experimenter we could only do this in a quite arbitrary fashion. However, most men would be satisfied with the value obtained, and would conclude that groups $A-F$ are not uniform.

It will already have occurred to the reader that with this test, as with U and T, there must often be tied ranks. These can be corrected for, using a process analogous to those which we described in Chapter 8. As before, if there is a group of equal scores, we give to each the mean of the ranks that would have been distributed among them had they been consecutively ordered. For each such group we form the quantity $C = t^3 - t$ where t is the number of tied scores in the group. If, in a total population of N observations there are more than one group of ties, we sum the several Cs to form ΣC, and use the modified statistic H', given by:

$$H' = \frac{H}{1 - \dfrac{\Sigma C}{(N^3 - N)}} \tag{9.3}$$

An example will readily demonstrate the use of this expression.

Let us suppose that a rat experimenter wishes to compare the behaviour under various experimental conditions of five groups of rats. He is, let us say, concerned with exploratory behaviour; and he wishes to satisfy himself that there is no major difference between his groups under control conditions. Each rat is in a cage, opening into a lighted arena; and the experimenter (or, more probably, his assistant) counts the number of "voluntary" excursions into the arena made by each rat in the course of a fixed period of time. The numbers are given in Table 9.2.

Table 9.2 Numbers of excursions made in a given period by the rats of several experimental groups (ranks of scores given in brackets)

	Group				
	A	B	C	D	E
Rat 1	9 (36½)	9 (36½)	9 (36½)	3 (10)	6 (21)
2	4 (14½)	1 (26½)	1 (2½)	7 (26½)	3 (10)
3	5 (18)	7 (26½)	7 (26½)	7 (26½)	3 (10)
4	9 (36½)	8 (31½)	2 (6)	3 (10)	7 (26½)
5	9 (36½)	6 (21)	1 (2½)	4 (14½)	3 (10)
6	1 (2½)	6 (21)	5 (18)	7 (26½)	3 (10)
7	4 (14½)	2 (6)	8 (31½)	4 (14½)	1 (2½)
8	9 (36½)	9 (36½)	9 (36½)	2 (6)	7 (26½)
T_i	195½	205½	160	134½	124½

The experimenter does not wish to run any serious risk of bias: i.e., he hopes that H_0 (the groups do not differ in spontaneous exploratory behaviour) is true, and he is resolved to proceed only if $\alpha > 0.1$.

We again use (9.2) to compute the quantity H, which is

$$H = \frac{12}{40.41} \left[\frac{(195.5)^2 + (205.5)^2 + (160)^2 + (134.5)^2 + (124.5)^2}{8} \right]$$
$$- 3.41 \doteq 127.7 - 123$$
$$= 4.7$$

However, the t terms are far from negligible. There are four scores of 1, three of 2, five of 3, four of 4, three of 5, three of 6, eight of 7, two of 8 and eight of 9. Thus the quantity ΣC is given by:

$$\Sigma C = (4^3 - 4) + (3^3 - 3) + (5^3 - 5) + (4^3 - 4) + (3^3 - 3) + (3^3 - 3)$$
$$+ (8^3 - 8) + (2^3 - 2) + (8^3 - 8)$$
$$= 1326$$

Hence (9.3) gives:

$$H' = \frac{H}{1 - \dfrac{1326}{40^3 - 40}} = \frac{4.7}{1 - \dfrac{1326}{63,960}} \doteq 4.8$$

This, then is the value of χ^2 on 4 d.f.; and, consulting Appendix Table D, we find $p \mid H_0) = 0.3$. This value is adequate even to still the doubts which might reasonably be felt when the number of ties is as large as it is here. Our happy experimenter is satisfied, and will continue to labour with his rats.

We have animadverted briefly, though persistently, to the difficulties in "matched-pair" designs: far greater are they when an attempt is made to match more than two cases. Nevertheless, the pressure to do so may be greater when we wish to compare many possibilities than when we wish to compare only two. The reasons for this are administrative rather than scientific. Supposing you wish to compare the effects of (say) six different treatments, and suppose that (say) eight individuals would make a minimum sized group. If you attempted to use separate groups, you would have to assemble at least forty-eight persons — more, preferably, to allow for unavoidable absences — with all the trouble that entails. If however, you gave all six treatments to every individual, only eight persons need be gathered, with large savings in time, trouble and money.

This being so, it is not surprising that techniques have been devised for taking advantage of this possibility, when it can be realised. That which concerns us here is the Friedman test (Friedman, 1937).

Suppose that N subjects are being tested for the effects of K different treatments. Since the number of subjects is necessarily the same for all treatments, the data are laid out in a table having K columns and N rows. The question under examination is whether or not the K treatments are equivalent; and clearly the null hypothesis is that they are: i.e., there is no difference.

Let the observations in each *row* be ranked from 1 to K, in ascending order of magnitude. If H_0 be true, then within each row the arrangement of ranks is merely a random permutation of the numbers $1 \ldots K$, each of the $K!$ possible arrangements being equally likely; and hence the $(K!)^N$ possible layouts of the table will also be equally likely. Let the rank total of the ith column be T_i; and consider, for any table, the quantity

$$S = \sum_{i=1}^{K} \left[T_i - \frac{N(K+1)}{2} \right]^2$$

Now, evidently the mean rank of any given row is $(K+1)/2$, and hence the average *column* total

$$= \frac{N(K+1)}{2} = \bar{T}$$

(say), always given H_0. Thus

$$S = \Sigma(T_i - \bar{T})^2 \tag{9.4}$$

The probability of obtaining an S as large as, or larger than, that actually obtained is clearly the number of tables actually yielding such values divided by $K!$. Such quantities have been tabulated for small values of N and K (Appendix Table I); but for larger values, though computers could produce the tables easily enough, they would become excessively bulky and inconvenient to use. Consequently a short cut has been sought and found.

Friedman was able to show that the quantity X^2, given by

$$X^2 = \frac{12S}{NK(K+1)} = \left[\frac{12}{NK(K+1)} \sum_{i=1}^{K} T_i^2 \right] - 3N(K+1) \tag{9.5}$$

is distributed as χ^2 with $(K-1)$ degrees of freedom. This enables us to use Appendix Table D in most cases.

An example seems called for at this stage.

Our friend from Tau Ceti has returned to Earth, armed with a Revulso-meter — which is a device for measuring the degree of revulsion induced in a human being by the close proximity of another one. (Negative scores are, of course, possible.) He obtains a set of eight human beings, and presents

each of them in turn with a journalist (J), a pimp (P), a television
personality (T), a clergyman (C), a blackmailer (B), and a lavatory
attendant (L). These specimens are presented to each subject in a different,
but otherwise random order; and the subjects' R-scores are noted in each
case.

The Cetian wishes to inquire whether these types differ in their
repulsiveness: H_0 is clearly that they do not. He rather expects that they
will, since the L is clearly a useful member of society carrying out a useful
function. His α therefore, is 0.083. (This is $\frac{1}{12}$; and the Cetians, in case you
do not know, base their counting on 12.)

The results of the experiment are given in Table 9.3.

Table 9.3 Revulsion scores recorded by subjects in the presence of six
individuals (ranks of scores are given in brackets)

	Test specimen					
	J	P	T	C	B	L
Subject 1	75 (3)	81 (4)	93 (6)	70 (2)	85 (5)	2 (1)
2	7 (2)	62 (3)	71 (6)	69 (5)	63 (4)	0 (1)
3	36 (3)	50 (4)	60 (6)	32 (2)	55 (5)	15 (1)
4	46 (3)	44 (1)	48 (5)	47 (4)	49 (6)	45 (2)
5	50 (2)	80 (5)	70 (4)	60 (3)	90 (6)	40 (1)
6	0 (1)	10 (4)	17 (5)	3 (3)	33 (6)	1 (2)
7	55 (3)	48 (2)	81 (5)	63 (4)	88 (6)	45 (1)
8	30 (3)	40 (4)	60 (5)	28 (2)	90 (6)	26 (1)
T_i	20	27	42	25	44	10

In order to determine X^2 we now substitute in (9.5), giving:

$$X^2 = \frac{12}{8 \cdot 6 \cdot 7} (400 + 729 + 1764 + 625 + 1936 + 100) - 3 \cdot 8 \cdot 7$$

$$= \frac{1}{28} \cdot 5554 - 168 \doteq 198.4 - 168$$

$$= 30.4$$

Consulting Appendix Table D for a χ^2 of 30.4 on $(6 - 1) = 5$ d.f., we find
that $p \,|\, H_0 < 0.001$. Our Cetian, therefore, considers himself justified in
supposing that there are large (and reasonable) differences. However, as we
shall see later, there are serious pitfalls which he may have overlooked.

A word of warning must be issued here to beginners. It may happen that an experimenter has assembled a batch of data looking like Table 9.2 or 9.3. There is often a strong temptation to compare only the two columns with the largest and smallest rank sums respectively and to use U or Wilcoxon's T for this purpose. The idea (we've all been tempted) is to avoid a lot of the arithmetic involved in (9.4) above. *This is invalid: don't do it.*

In any population there has to be a largest and smallest member. If you try the simplification described, you *may* be only looking at the extremes of a continuous population; and you can end up by fooling yourself into thinking that the two samples come from different populations when you only have the "tails" of one. It is only doubtfully valid to do this after H or X has been used to reject H_0 — by which time the exercise is usually uncalled for.

All the statistics we have talked about in the last chapter and this, U, T, H and X^2 are essentially tests for differences in *location*. That is: they examine whether samples differ in their *median*. But as we saw quite early on, there are other differences which are important, especially differences of spread. There are a number of tests for this sort of difference, although all of them are subject to certain limitations.

These limitations are inherent in the ranking process: suppose that the smallest score in a given batch has the value -150, and the second smallest -3, these will have ranks 1 and 2 respectively; but so also would scores of -4 and -3 if these were the smallest and second smallest. Thus, inevitably, we lose information. Also, one of the handiest tests for "spread", the Siegel–Tukey test (Siegel and Tukey, 1960), can only be used when the populations under consideration do *not* differ significantly in median. So long as this is kept in mind, however, it can be very useful. The mathematical formulation of this test is formally the same as that of the U-test, the constructors' ingenuity being displayed in altering the ranking procedure to answer a different question.

Consider the following example. An inspector of schools is doubtful whether a newly proposed technique for streaming pupils is a good one. He arranges for a standardized test to be given to a batch of children who have been assigned to a particular stream by method X, and the same test to be given to another batch, assigned to the same stream by method Y. The scores obtained are shown in Table 9.4.

It is evident upon inspection that the two sets of scores have the same median: 50. However, while the scores in the Y group have a range of 29, those in the X group have a range of 78. This looks very impressive, but what is it worth as evidence?

In the Siegel–Tukey test we proceed to rank the two samples taken together, but not in the way we have become used to. For this test we rank

Tablo 9.4a Scoros obtoinod by children selected by two different methods

X group scores	Y group scores
86	66
80	59
68	54
58	50
42	45
29	40
21	37
8	

Table 9.4b The above data ranked for the Siegel–Tukey test

X ranks	Y ranks
2	8
4	10
6	14
12	15
11	13
5	9
3	7
1	

from the extremes inwards. Thus we assign rank 1 to the smallest score, rank 2 to the largest, rank 3 to the second smallest, rank 4 to the second largest and so on. The ranking in our example is given in Table 9.4b.

We must go back and ask a couple of important questions here. First, what sort of question is the inspector asking? If (as we may suppose) he starts with the suspicion that method X is inferior to − i.e., yields a greater spread than − test Y, it is clearly a one-tailed question. Second, what of α? This is necessarily harder to determine, so let us assume the traditional value of 0.05.

We note that the rank sum for the smaller group (i.e. that of seven subjects) is 76. We then consult Appendix Table J and find that α is satisfied; and so is the inspector of schools.

The reader can easily satisfy himself that this test is not valid unless the means are not importantly different, by considering a particular case. Suppose that there are two samples differing as much as you please in

range, but with *no overlap at all*. Then the ranking method recommended
by Siegel and Tukey will assign successive ranks alternately to the two
samples: one would have ranks 1, 3, 5 . . . , the other ranks 2, 4, 6. . . .
Thus no difference in spread *can* emerge.

In this chapter we have been looking at three techniques, based on the
method of randomization, which enable us to tackle more complex
questions than were raised in Chapter 8.

If we wish to compare K different treatments, and have K independent
sample groups upon which to try them out, we may examine possible
differences in median using the *Kruskal—Wallis* test. To do this we rank all
the N readings of the several groups together, assigning rank 1 to the
smallest score, and rank N to the largest. If (as is usual) the data are set out
in a table with one column to each condition, we find the several column
totals T_i, and hence compute the quantity H using (9.2).

If N is small, we now consult Appendix Table H for significance level. If
N is large, on the other hand, we regard H as a value of χ^2 with $(K - 1)$ d.f.,
and consult Appendix Table D for the same end. If there are tied scores,
but not too many, we compute the correction terms C, and hence H', using
(9.3). The tables are then consulted as before.

If we have managed, by some remarkable skill, to obtain *matched* scores
for the K conditions, we can use *Friedman*'s test. If we lay out our scores
in the usual manner once again, we will have K columns and N rows, where
each row is a set of matched scores. We now rank the scores in each *row*,
and add the ranks so obtained in each *column*, to obtain the column
totals T_i. These we can substitute in (9.4), to obtain the quantity S if N
and K are small; and Appendix Table I may be used to find $p(H_0)$. If N
and K are *not* small we use the approximation given by (9.5) for $X^2 = \chi^2$
with $(K - 1)$ d.f.

If we have two samples *which we know to be more or less indistinguishable in median*, but which we suspect to differ in *spread*, this possibility
may be tested using the Siegel—Tukey test.

Here, the two samples are taken together, but are ranked, not
monotonically with respect to magnitude, but from the extremes of
magnitude towards the centre. The simple total of the ranks so obtained
is found, and Appendix Table J used to check its probability.

Exercises

1. Aristotle maintained that persons dwelling south of Hellas were (com-
paratively speaking) wanting in spirit. His erstwhile and repugnant pupil
Alexander of Macedon sent to Athens some groups of slaves he had
acquired during the course of his campaigns, only one group consisting of
Hellenes. By this time Aristotle's other (and now forgotten) pupil

Eisenkides had produced a pencil-and-paper (or rather reed and wax) test for spiritedness. This test was given to some of these groups, with results given in Table 9.E1. Comment.

Table 9.E1   Spiritedness scores from sundry groups of slaves

	Group					
	Hellenes	Bactrans	Sogdians	Indians	Medes	Egyptians
Score	96	54	18	19	74	1
	32	98	27	48	50	97
	33	58	85	26	9	42
	67	84	20	3	95	2
	99	70	15	29	41	92
	73	—	90	39	—	88

2. Prove that $(9.1) = (9.2)$. (Go on, it's not all that difficult!)

3. Recent disasters (or successes, depending on point of view) with induced fertility have produced some numbers of identical triplets. A group of afflicted (or happy) parents has cooperated to test the results of certain diet additives known as X, Y and Z, upon the height of the children: one triplet of each set having one of the additives. The table gives the heights of various sets of triplets in centimetres, on their ninth birthdays. Is there evidence for the differential influence of the additives?

Table 9.E3   Heights (cm) of triplets dosed with substances X, Y, Z

		Additive		
		X	Y	Z
Triplet set	1	149	150	151
	2	147	148	149
	3	140	142	139
	4	136	134	135
	5	135	137	138
	6	139	140	141
	7	140	139	141
	8	142	141	140
	9	148	146	145
	10	148	149	150

4. An advertising concern is interested in persuading the Great British Public to accept a number of quite mendacious propositions. Five Television Personalities (A–E) have readily agreed to prostitute themselves to this end; and groups of potential mugs have been collected for a trial. Each group is told twelve flat lies by one of the TVPs; and the number believed by each subject recorded. The results of the trial are given in Table 9.E4. Does it provide evidence upon which to pick the most successful TVP?

Table 9.E4   Numbers of lies believed by separate groups of subjects, told by five TVPs

		Liars				
		A	B	C	D	E
Subject	1	9	5	3	5	9
	2	9	7	7	7	9
	3	9	7	9	3	12
	4	3	7	6	4	7
	5	6	5	1	2	5
	6	4	9	4	8	3
	7	7	8	3	4	8
	8	1	2	1	1	4
	9	7	–	6	9	9
	10	3	–	–	10	10

References

Friedman, M. "The use of ranks to avoid the assumption of normality implicit in the analysis of variance", *J. Amer. Statist. As.*, 1937, **32**, 675–701.

Kruskal, W. H. and Wallis, W. A. "Use of ranks in one-criterion variance analysis", *Ibid.*, 1952, **47**, 583–621 and 1953, **48**, 907–11.

Siegel, S. and Tukey, J. W. "A nonparametric sum of ranks procedure for relative spread in unpaired samples", *Ibid.*, 1960, **55**, 429–55.

10

If it is bell-shaped

In the last few chapters we have been discussing methods and tests which made few demands upon the precise form of the distribution of the data being studied. Such methods are, therefore, known as "distribution free" or "non-parametric". (Strictly speaking, these terms are not equivalent, although by usage they have virtually become so.) Because of this fine parsimony of assumption, they are usable in almost any situation. However, a price is paid for this versatility: a method which can be used with *any* distribution is inevitably less powerful in dealing with some specific distribution than a test tailor-made for that specific case.

Since a great deal of theoretical work has been done on the Gaussian distribution, and since a number of real-life distributions approach it closely, there are many tests which may be used with advantage when we find that our data have that form. In the behavioural sciences, the most commonly used test which is based upon the Gaussian distribution is the *t*-test, often employed with gay abandon and not the slightest check to see whether its use is truly valid. (*Peccavi.*)

The *t*-test was devised by one W. S. Gosset (who worked for the Guinness concern, whose products I hope he sampled carefully), who always employed the pseudonym of "Student" on his mathematical papers; consequently it is often referred to as "Student's *t*". The actual derivation employs rather more advanced techniques than we are discussing in this book, so we will discuss the use and limitations of the test without describing its formulation.

Let us suppose that we have a sample of size n, whose mean value is $\bar{x}$ and we wish to know whether it could have come from a Gaussian population whose mean is μ and whose s.d. is σ. That is to say, we wish to know whether the difference between $\bar{x}$ and μ could reasonably be ascribed to mere sampling error. "Student" suggested that, for this purpose, we use the statistic "t" defined by:

$$t = \frac{\text{difference in means}}{\text{standard error of mean}}$$

$$= \frac{|\mu - \bar{x}|}{(\sigma/\sqrt{n})} \quad \text{(see Chapter 5)}$$

$$= \frac{|\mu - \bar{x}|\sqrt{n}}{\sigma} \tag{10.1}$$

If the quantities are known, the arithmetic is straightforward; but often σ is not known. For example, we may have a theory which predicts a value for μ; but it is much rarer to have one which predicts σ. We are therefore compelled to approach the question backwards; if we assume that the sample is indeed from the parent population, we can use the sample s.d. ($= s$) to estimate σ; and as we saw (Chapter 5, op. cit.) the best estimate we can obtain is

$$\sigma = s = \left[\frac{1}{n-1} \sum_{i=1}^{n} (\bar{x} - x_i)^2 \right]^{\frac{1}{2}}$$

Substituting this in (10.1) we obtain

$$t = \frac{|\mu - \bar{x}|\sqrt{(n)}}{s} \tag{10.2}$$

Let us consider an example.

On the back of the "Highway Code" there are a series of statements implying that the probable reaction time to road emergencies is 0.625 secs. A mock-up of a driver's position has been prepared, and subjects sit in it and watch a film of a road moving towards them. At one point in the film, a child jumps into the road, and the time taken to initiate an emergency stop is measured. The results for ten subjects are given in Table 10.1.

Table 10.1   Reaction times of ten subjects to a simulated emergency

	Subject									
	1	2	3	4	5	6	7	8	9	10
R.T. (secs)	0.250	0.343	0.375	0.430	0.450	0.465	0.490	0.550	0.577	0.680

We may readily compute that the sample mean $\bar{x}$ is 0.461. If we suppose that the sample s.d. is our best pointer to the population s.d. (and we have little other choice if we are to proceed at all), then since $n = 10$, the best s.d. of the sample is 0.117; and hence

$$t = \frac{|0.625-0.461|(10)^{\frac{1}{2}}}{0.117} = \frac{0.164 \cdot 3.162}{0.117}$$

$$\doteq 4.4$$

In order to consult Appendix Table K, we must now settle the number of degrees of freedom available in this case: it is one less than the number of items in the sample, i.e. $(10 - 1) = 9$. Our H_0 is evidently that the sample could have come from a population whose mean is 0.625 with s.d. as computed; and as we have no strong feelings about the Highway Code we might be prepared to settle for an α of 0.10.

Alas, consulting the table we see that even this slack criterion is not satisfied. $p(H_0)$ lies between 0.1 and 0.01. It would seem that 0.625 sec is an over-estimate of the relevant reaction time.

Thus the t-test when testing a sample mean; but how do we proceed when comparing the means of two independent samples?

The *assumption* we make here is that both samples are drawn from Gaussian populations of equal s.d.; and H_0 is that the two independent samples come from the *same* population: i.e., if we call the items in one sample x_i and in the other y_i, with our usual notation H_0 is that $\bar{x} = \bar{y}$. In this case the value of t is:

$$t = \frac{\text{difference of means}}{\text{standard error of difference of means}}$$

Here we have to use the pool of items n_1 xs (say) and n_2 ys to form an estimate of the variance of the supposed parent population. Denoting the sample s.d.s by S_x and S_y, the best estimate is

$$S^2 = \frac{n_1 S_x^2 + n_2 S_y^2}{n_1 + n_2 - 2}$$

and the standard error of the difference of the means is

$$S \sqrt{\left(\frac{1}{n_1} + \frac{1}{n_2}\right)}$$

whence

$$t = \frac{|\bar{x} - \bar{y}|}{\sqrt{\left[\left(\frac{n_1 S_x^2 + n_2 S_y^2}{n_1 + n_2 - 2}\right)\left(\frac{1}{n} + \frac{1}{n_2}\right)\right]}} \cdots \tag{10.3}$$

The result having

$$(n_1 + n_2 - 2) \text{ d.f.}$$

We can now consider an example of the use of this form of the test.

The Serious Minded Cannibal, whom we encountered a few chapters ago, wishes to decide whether it were more profitable to seek the company of Catholic missionaries or Protestant ones. He consults his records for the drawn weight of specimens he has already tried, and draws up his data in Table 10.2.

Table 10.2   Drawn weights (kilos) of Catholic (C) and Protestant (P) missionaries

C	P
37	38
43	41
45	44
45	46
48	47
54	50
	56

We see that

$$n_1 = 6, \qquad n_2 = 7;$$
$$\bar{x} = 45.3, \qquad \bar{y} = 46.0$$

It does not seem very likely that this small difference amounts to much; but the SMC is always out for the last gram of protein, and sets α at 0.1.

We speedily find that

$$S_x^2 = 157$$
$$S_y^2 = 210$$

Hence substituting (10.3) gives:

$$t = \frac{0.7}{\sqrt{\left[\left(\dfrac{6.157 + 7.210}{6 + 7 - 2}\right)\left(\dfrac{1}{6} + \dfrac{1}{7}\right)\right]}}$$

$$\doteq \frac{0.7}{\sqrt{67.9}}$$

$$= 0.085$$

A swift glance at Appendix Table K dispels any hope of rejecting H_0 upon the chosen α: the SMC might as well toss a penny — or a bone.

The reader might well have passed over the assumptions for the use of t. Since it is assumed that the two samples are of equal variance, were it not desirable to see whether the assumption is met? Yes indeed, and there is a very simple and straightforward statistic for testing it. This is the Variance Ratio or F-test. (F in honour of the great and irascible statistician R. A. Fisher, who studied the fundamental z distribution upon which the test is based.)

Once again, a complete exposition of the derivation of this test is outside the scope of this book, and only the procedure will be given. This is simplicity itself. In general, two samples will yield different sample variances. Let these be S_1^2 and S_2^2, where $S_1^2 > S_2^2$. Then

$$F = \frac{S_1^2}{S_2^2}$$

In the example we have just looked at,

$$F = \frac{210}{157} \doteq 1.34$$

We consult Appendix Table L to find the associated probability of F under H_0, i.e., the probability of finding as large an F, given that there is indeed no difference between the variances.

Unfortunately, this probability is a function of the degrees of freedom associated with each sample. Consequently, the tables are laid out only for fixed values of α; the traditional ones of 0.05 and 0.01. To use the table, it is entered at the column corresponding to the number of d.f. associated with S_1^2 (6 in this case) and the row corresponding to the number of d.f. associated with S_2^2 (5 here). If the value of F is *larger* than the value at this cell in the table, then $p|H_0$ is *less* than the α for which the table is constructed. Our table shows that we cannot reject the present H_0 at $\alpha = 0.05$.

Now, if the reader has mixed much with psychologists and other low company, he may have heard talk of "robustness", and be inclined to repeat the hoary tag "Well, it doesn't matter, anyway, since it is a robust test." It must be emphasized that, as it stands, that remark is *meaningless*. "Robustness", in a test, means that it remains valid if one or other of the assumptions upon which the test is based is not met. *No test is, or can be, robust against the violation of all the underlying assumptions, or against extreme violations of any of them.*

What then of the t-test? The assumptions made, as you will recall, are (i) the samples are Gaussian and, (ii) of the same variance. So long as (i)

holds, the reader may take it that the test may be used so long as F does
not reject the hypothesis of equal variances, certainly at the 0.05 level, and
probably at the 0.01 level. Unfortunately, there is a circularity here; for
the F test also depends upon (i) holding good. How about assumption (i)
then? Here again, one hears a lot of nonsense talked; for distributions can
depart from the Gaussian form in more than one way. The importance of
such departures for t has been very extensively studied. The conclusions
generally reached are that, so long as symmetry is preserved, quite large
departures in mesokurtosis can be accepted; but that skewness is much
more serious. Put into common speech: it doesn't matter too much if the
distribution is pretty flat, so long as it isn't lop-sided. (See Fig. 10.1.)

Fig. 10.1

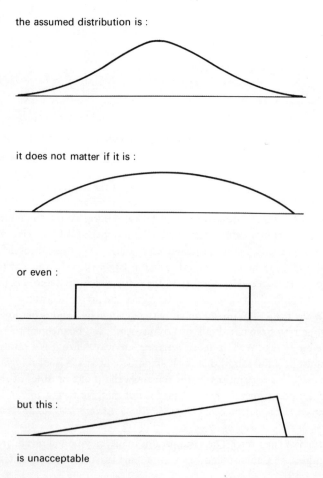

How can the user satisfy himself then that (i) is adequately satisfied? Unfortunately, this can be quite difficult in practice. There are methods available, but they require sample sizes of at least thirty; and if you have as many as that you could use χ^2 to test for deviations from a Gaussian distribution (see, e.g., Daniel and Wood, 1971). Quite often, the best thing to do is to plot a rough histogram and use your commonsense: if it looks notably skew, eschew t.

The reader may well be wondering, at this point, why, if all these tests and precautions are desirable, anyone should want to use t at all. The reason is simply this: that, *when its assumptions are satisfied, t* is very powerful. In other words, the probability of rejecting H_0 when it is false (i.e., the probability of correctly rejecting H_0) is greater for the t-test than for the distribution-free tests we have looked at. Admittedly, the difference is not great: if t validly rejects H_0, the probability that, e.g., Wilcoxon's T will do so is approximately 0.86 (this, of course, assumes that the underlying distributions are indeed Gaussian), but still, this could mean failing to reject H_0 once in seven experiments. (There are those — and I have much sympathy with them — who say that if the results are so close to unacceptability that the difference between two alternative test procedures matters, you are dealing with something essentially trivial. The reader must decide for himself.)

At this point we should go on to consider analysis of variance and the somewhat elaborate experimental designs which it is designed to cope with, and the pitfalls inherent in, these designs for the human scientist. However, so complex are some of the questions raised by these methods that they really deserve a chapter to themselves.

In this chapter we have looked at a powerful method for determining whether two samples can be regarded as differing in mean, which is valid when both distributions are roughly symmetrical and of similar variance.

If there are two samples to compare, use expression (10.3). If we have to compare a sample mean with some previously determined value, use (10.2).

Exercises

1. There has been a wave of murders performed with the aid of a Mysterious South American Arrow Poison. Two toxicologists, A and B, examine the nine corpses available, and for each corpse produce an estimate of the amount of poison which had been introduced into the victim before death. The table gives the two sets of estimates. Has the defending council grounds for declaring that the two experts differ too much to be trusted?

	Corpse								
	1	2	3	4	5	6	7	8	9
Estimate A (mg)	7.50	8.60	8.80	9.40	8.60	9.70	7.90	6.70	9.90
Estimate B (mg)	7.50	9.60	10.30	9.20	8.70	8.10	8.20	6.30	9.00

2. A factory uses two machines for making cakes. All cakes are supposed
to be of the same mean weight; but an inspector entertains doubts about
the process, and weighs two randomly selected batches of ten, one batch
from each machine. The table gives the readings he obtained: were his
doubts justified?

	Cake No.									
	1	2	3	4	5	6	7	8	9	10
Batch A (mass in g)	240	236	240	244	243	238	239	239	246	242
Batch B (mass in g)	246	240	245	245	249	246	243	248	246	247

3. A manufacturer of motor vehicles claims that one of his models con-
sumes 4.50 litres of petrol per 50 km. A salesman of great (indeed, unique)
honesty puts precisely 4.50 litres into the tanks of each of the eleven
specimens in his salesroom, and drives each of them round a measured
circular 1-km track at recommended optimum cruising speed until the
juice runs out. The table gives the distances covered: what does the
salesman make of the manufacturers claim?

Car No.	Distance covered (km)	Car No.	Distance covered (km)
1	50.60	7	48.13
2	45.72	8	47.22
3	49.58	9	46.73
4	46.45	10	47.37
5	48.57	11	44.82
6	46.63		

4. An experimenter is measuring the response times of rats to stimuli of
two different kinds. The rat stands upon a grid, and when a current is
made, a solenoid presses a rod into the rat's behind. In group A, the rod is
blunt ended; in group B it ends in an acutely pointed double prong. The

time between the onset of current and the rat's jump is timed to the nearest millisecond, the results being given in the table. Do the results justify a belief that the stimuli produce different response times?

Group A		Group B	
Rat No.	R.T. (m.s.)	Rat No.	R.T. (m.s.)
1	131	1	133
2	133	2	145
3	136	3	152
4	142	4	166
5	143	5	171
6	145	6	177
7	148	7	177
8	150	8	182
9	152	9	184
10	164	10	186
11	166	11	186
12	177	12	191
13	183		

Reference

Daniel, C. and Wood, F. S. *Fitting Equations to Data*. New York, Wiley, 1971.

11

Analysis of variance (or: what not to do and how to do it)

The analysis of variance was originally devised to cope with experimental designs which were themselves devised for agricultural research. For the purposes originally in mind they are brilliantly successful; but, unfortunately, they are often applied to experiments with human subjects, where they have some inherent weaknesses, which are too readily overlooked.

When we have spoken about the "variance" of a population heretofore, we have always been thinking of the overall variance of a single, supposedly more-or-less homogeneous population. However, it often happens that, within some large overall group, there is more than one source of variability. For example, if we measured the heights of the population of a town, there would be variability due to age and sex, as well as to the genetic variation between individuals of the same age and sex.

If we proceed from human heights to consider (say) crop yields — the topic of the utmost importance to agriculturalists — it will be evident that yields (measured in kilos/hectare) will vary with variety of seed, type of soil, type and quantity of manure and so on. It is important to find the principal sources of variability, and to distinguish the more from the less important. Roughly, then, the paradigm of the agricultural experiment is this: a number of blocks of land, each of which is small enough to be regarded as reasonably homogeneous in the fertility of the soil, is further sub-divided into plots. On each plot a different variety of seed is grown, or a different type of manure is applied. The object of the whole exercise is to see whether one variety, or one sort of manure yields better crops, or higher resistance to disease, or whatever question is of interest.

This is very effective; and the translation into the human sciences is effected by substituting subjects for blocks (there are those who think that this is highly appropriate) tests for varieties of seed or manure, and presentations of tests for plots.

There are many different ways of conducting experiments based upon this general notion; but the most common are what are called "randomized blocks" and "Latin Square" designs. Some account of both, and of the purposes they are intended to serve, is in order.

In a "randomized blocks" design, treatments are allotted randomly, not to the blocks, but to the plots within each block. Thus, supposing that all members of a group of subjects were to receive tests A, B and C . . . N, the N tests would be administered to each subject in a different random order. It is argued, with considerable apparent plausibility, that presenting the tests in random order cancels out the effects which learning or fatigue might have, in improving or degrading performance on those tests which were administered later.

The "Latin Square" design is a refinement of this. Let us suppose that we were satisfied that the randomized blocks design had eliminated the sources of error we have mentioned, but also wanted to check whether the hour at which the subject had taken the test was also important (it might be). This concern can be handled using a Latin Square.

A Latin Square is an array of letters like this:

A	B	C	D
C	A	D	B
D	C	B	A
B	D	A	C

Its defining characteristics are that (i) there are as many rows, or columns, as there are different letters, and (ii) that each letter occurs exactly once in each row and column, but always in a different position in each. Now we have supposed that we are concerned with three sources of variability: the variability due to the different tests or treatments being applied (known as between-groups variance), the variability due to differences between individual subjects (known as within-group variance) and that due to time of presentation.

Let us now interpret our Latin Square thus: let each letter represent a particular test, let each row represent the order of presentation of tests to a particular subject and let each column represent a time of presentation. If we randomly assign letters to tests, rows to subjects and columns to times, we may claim to have balanced out the effects of unwanted variables. (We will return to this very important point later.)

An obvious apparent limitation of the design is that it is always $n \times n$: i.e., it can only allocate as many subjects and times as there are tests. This is less trouble than it looks, however; for it is always possible to replicate over successive groups, or (sometimes better) to use different Latin Squares for each group of subjects — and we need large numbers for validity in the human sciences.

Thus (for the moment) for a couple of popular experimental designs.

What do we do with the data we get? It is time to backtrack to the randomized blocks experimental design and see how we play with our numbers.

Let us suppose that three different designs of keyboard for a desk calculator are being compared. A randomized blocks design is employed, each keyboard being assigned to each subject in a random order. There are, let us say, six subjects, and the score under each condition is the number of sums, of (let us assume) equal difficulty, correctly carried out in a pre-set time. The results are laid out in Table 11.1.

Table 11.1   Results of a keyboard comparison experiment

		Keyboard			
		A	B	C	Subject's total
Subject	1	8	9	10	27
	2	7	8	12	27
	3	6	7	11	24
	4	6	7	8	21
	5	8	9	10	27
	6	7	8	9	24
Keyboard's total		42	48	60	Overall mean 8.33

The table presents a population of eighteen values, drawn from six subjects under three conditions. When we perform an analysis of variance, we *assume* that each of these eighteen terms consists of four components: the overall population mean, a term related to the particular subject, a term related to the particular condition (Keyboard in this case) and an "error" or random variability term. It is further assumed — and, for the moment, let us go along with this — that these four terms are *additive* and *independent*.

Before proceeding, let us introduce a little clarifying notation. Suppose that there are n subjects and m tests — i.e., n rows and m columns in our table. Let us then denote the score of the ith subject on the jth test as x_{ij} (where, of course, $i = 1 \ldots n$ and $j = 1 \ldots m$). Since the results table is a rectangular array, the total number of entries will be $nm = N$ (say).

The measure of overall variability we need is called the total sum of squares, which we will write as "total ss". It is

$$\Sigma\Sigma \, x_{ij}^2 - \text{C.F.}$$

[The double Σ signifies that the term following is summed over all ij pairs.]
C.F. being a correction factor given by

$$\frac{(\Sigma\Sigma\ x_{ij})^2}{mn}$$

so that:

$$\text{total } ss = \Sigma\Sigma\ x_{ij}^2 - \frac{(\Sigma\Sigma\ x_{ij})^2}{mn} \tag{11.1}$$

We now encounter once again the term "degrees of freedom". Since, if $\bar{x}$ is known, $N - 1$ readings leave no uncertainty about the Nth, we say that $(N - 1)$ d.f. are associated with total ss. In the present example, $N = 18$, and substituting in (11.1) gives:

$$\text{total } ss = 1296 - \frac{(150)^2}{18} = 46 \quad \text{with 17 d.f.}$$

We must now seek for the components of this overall variability. Since we have assumed that these components derive from test variability, subject variability and random (residual) variability, we require measures for each of these. Just as for the total variability, the *sum of squares* associated with each of these sources is the appropriate measure. We have:

$$ss \text{ for tests} = \sum_{j=1}^{m} \frac{\left(\sum_{i=1}^{n} x_{ij}\right)^2}{n} - \text{C.F.}$$

$$= \Sigma_j \frac{(\Sigma_i\ x_{ij})^2}{n} - \frac{(\Sigma\Sigma\ x_{ij})^2}{mn} \tag{11.2}$$

Obviously, for a given j, $\Sigma_i\ x_{ij}$ is a *column total*, so 11.2 is a shorthand for:

$$ss \text{ for tests} = \text{sum of all } \frac{(\text{column totals})^2}{\text{number of rows}} - \text{C.F.}$$

which, in our example, comes to:

$$\frac{42^2}{6} + \frac{48^2}{6} + \frac{60^2}{6} - \frac{150^2}{18} = 28$$

and as there are m tests (columns), there must be $(m - 1)$ d.f. associated with this term or, in this instance:

$$3 - 1 = 2 \text{ d.f.}$$

Similarly, for the variability between subjects we require the *sum of squares for subjects*. This straightforwardly enough is

$$ss \text{ for subjects} = \text{sum of all } \frac{(\text{row totals})^2}{\text{number of columns}} - \text{C.F.}$$

or, formally

$$= \Sigma_i \frac{(\Sigma_j x_{ij})^2}{m} - \frac{(\Sigma\Sigma x_{ij})^2}{mn} \tag{11.3}$$

And, naturally, if there are n subjects (rows) there must be $n - 1$ d.f. associated with this term. In our case:

$$ss \text{ for subjects} = \frac{27^2}{3} + \frac{27^2}{3} + \frac{24^2}{3} + \frac{21^2}{3} + \frac{27^2}{3} + \frac{24^2}{3} - \frac{150^2}{18}$$

$$= 10 \quad \text{with } 6 - 1 = 5 \text{ d.f.}$$

Since we have assumed that all the variability comes from three sources, a term related to the third and last — the *residual sum of squares*, denoted by R can be simply determined by

$$R = \text{total } ss - (ss \text{ for tests} + ss \text{ for subjects}) \tag{11.4}$$

and, since the total number of degrees of freedom must be accounted for, the number associated with the residual sum of squares is:

$$(N - 1) - [(m - 1) + (n - 1)]$$

which, in our example is $17 - (5 + 2) = 10$.

Further, substituting in (11.4) gives:

$$R = 46 - (28 + 10) = 8.$$

From each sum of squares term we can now produce a *variance estimate*, which is the sum of squares divided by the associated number of d.f. This has been done in Table 11.2, which shows how a completed analysis of variance table should look.

Table 11.2 Analysis of variance

Source of variation	Sum of squares	d.f.	Variance estimate	Variance ratio
Between Ss	10	5	2	2.5
Between conditions	28	2	14	17.5
Residual	8	10	0.8	—
Total	46	17	2.7	—

Now that we have variance estimates for the subjects, for the tests and for the residual, how should we use them in order to find which sources of variation are important? We can answer this by considering what our null hypothesis is. If there were indeed no real difference between conditions or subjects, the only differences which would appear would be those due to error: i.e., to the residual term. Our H_0, therefore, is that the several sources of variance are not importantly larger than the residual term R. We therefore use as our measure of the importance of each source of variability the *Variance Ratio*: i.e., the ratio of its variance estimate to the variance estimate of the residual term. The resultant V.R. is strictly equivalent to the F ratio we looked at in Chapter 10; and we consult Appendix Table L to examine its probability.

From Table 11.2, we see that, for differences between conditions (i.e., between keyboards), $F = 17.5$ with 2 and 10 d.f. Entering the tables, we find that $p \mid H_0 < 0.001$. Similarly, for differences between subjects, $F = 2.5$ with 5 and 10 d.f.; and we find that $p(H_0) > 0.05$.

In this case we did not decide upon α; but any man who declined to believe that there really is some advantage in keyboard C were hard to convince indeed; while the subjects do look rather much of a muchness.

Analysis of variance with a Latin Square design is essentially similar; but can become a little burdensome arithmetically; so we will create a rather simple little example for ourselves, using rather fewer subjects than any self-respecting experimenter should.

Our experimenter has devised three tests for verbal ability. All are supposedly of equal difficulty, being of the nature of acrostics; but the solution of one consists of rude words, one of neutral words, and one of pleasing words. He believes that this might make a difference to the time taken to solve them. He also believes that the time of day might affect the results; and having only three tests, produces 3 × 3 Latin Square design thus:

A   B   C
C   A   B
B   C   A

The letters were randomly allocated to tests, the columns to time of day (morning, afternoon and evening), and the rows to the subjects in time and order of presentation: thus subject 1 had test A in the morning, B in the afternoon, and C in the evening. Subjects' scores are the times (minutes) taken to solve the tests. Altogether there must be nine such scores, so the whole results population will have 8 d.f.

We are asking three questions: is there important variation between *tests*? Is there important variation between *times*? And is there important variation between *subjects*? Ideally we would like to lay out the data in three dimensions, but failing that we must use the layout of Table 11.3.

Table 11.3   Results of a study conducted with a Latin Square design

Times (iii)	Subjects (i)				Tests (ii)		
	1	2	3	Total	A	B	C
Morning	A4	B11	C8	23	4	11	8
Afternoon	C3	A12	B10	25	12	10	3
Evening	B4	C10	A12	26	12	4	10
Total	11	33	30	74	28	25	21

Once again, our assumption is that the overall variability is due to simply additive factors: this time there is one more. These are variability due to tests, to subjects, to times and to random (residual) variance. Again, also, the appropriate sum of squares term is used to measure this variability.

A little more notation may help.

It is inherent in the design that the number of subjects = number of tests = number of treatments (which are times, in this case) = n, and that there are (as we remarked above) n^2 observations altogether. Then:

let A_i be the total scores of the ith *subject* $(1 < i < n)$

(i.e., the several A_is are the column totals on the left of Table 11.3)

let B_j be the total scores obtained in the jth *test* $(1 < j < n)$

(i.e., the several B_js are the column totals on the right of Table 11.3)

let C_k be the total scores obtained under the kth *treatment* $(1 < k < n)$

(i.e., the several C_ks are the row totals in the middle of Table 11.3) further let

$$S = \Sigma\Sigma \, x_{ij}^2$$

while

$$\text{C.F.} = \frac{(\Sigma\Sigma \, x_{ij})^2}{n^2}$$

We can now write down the sums of squares and number of degrees of freedom associated with each source of variability.

$$\left.\begin{array}{ll}
\text{Total } ss & = S - \text{C.F. with } n^2 - 1 \text{ d.f.} \\[4pt]
ss \text{ for subjects} & = \Sigma_i \dfrac{A_i^2}{n} - \text{C.F. with } n - 1 \text{ d.f} \\[4pt]
ss \text{ for tests} & = \Sigma_j \dfrac{B_j^2}{n} - \text{C.F. with } n - 1 \text{ d.f.} \\[4pt]
ss \text{ for treatments} & = \Sigma_k \dfrac{C_k^2}{n} - \text{C.F. with } n - 1 \text{ d.f.}
\end{array}\right\} \qquad (11.5)$$

Residual ss is obtained by subtraction, and has $n^2 - 3n + 2$ d.f. And, in each case, the variance estimate for each factor is the sum of squares divided by the number of d.f.

We are now in a position to tackle our data. Substituting, we have:

$$\text{Total } ss = 4^2 + 11^2 + 8^2 + 3^2 + 12^2 + 10^2 + 4^2 + 10^2 + 12^2 - \frac{74^2}{9}$$
$$= 105.6 \ (8 \text{ d.f.})$$

$$ss \text{ for subjects} = \frac{11^2}{3} + \frac{33^2}{3} + \frac{30^2}{3} - \frac{74^2}{9} \qquad = \ 94.9 \ (2 \text{ d.f.})$$

$$ss \text{ for tests} = \frac{28^2}{3} + \frac{25^2}{3} + \frac{21^2}{3} - \frac{74^2}{9} \qquad = \quad 8.3 \ (2 \text{ d.f.})$$

$$ss \text{ for treatments (times)} = \frac{23^2}{3} + \frac{25^2}{3} + \frac{26^2}{3} - \frac{74^2}{9} = \quad 1.6 \ (2 \text{ d.f.})$$

$$\text{whence residual } ss = 105.6 - (94.9 + 8.3 + 1.6) \ = \quad 0.8 \ (2 \text{ d.f.})$$

We can now draw up the Analysis of Variance table for the data (Table 11.4).

Table 11.4   Analysis of variance of data in Table 11.3

Source of variation	d.f.	Sum of squares	Variance estimate	Variance ratio
Between tests	2	8.3	4.15	10.4
Between subjects	2	94.9	47.45	118.6
Between times	2	1.6	0.80	2.0
Residual	2	0.8	0.40	—

Since our experimenter expected marked differences between tests and times, he might have set the α for these sources at 0.05; while he cautiously put an α of 0.01 for differences between subjects.

Consulting Appendix Table L, however, we find that the experimenter's pets fail to satisfy his α values, while the between-subject term does. In this test, we conclude, difference between subjects is more important than the nature of the material or the time of day.

It is time to sit back and think.

On the face of it, analysis of variance is the answer to a behavioural scientist's prayer, at least when applied to an experiment based upon a suitably balanced design. Unwanted order effects are supposedly eliminated; more than one factor can be tackled at a time, and their relative importance compared; and the actual numerical effort required, though tedious, is not difficult. What is the catch? Unfortunately, there are at least two.

The most obvious is in fact the least troublesome. This is the assumption (see above) that the several factors under consideration are *independent* and *additive*. This is a rare case in the human sciences. It is notorious that, to take an obvious example, though there may well be real differences between (say) types of intellectual test and between students, these are not independent. Student A prefers tests of type X, while student B prefers those of type Y: in other words the factors interact.

However, this, by itself, is not an insuperable problem; for there are other — admittedly somewhat more complex — forms of analysis which can cope with and estimate interaction terms.

The real bother is this: the assumption that random or balanced orders of presentation eliminate inter-condition influences *is not found to be valid for human subjects*. Human beings can learn so subtly, and adapt so swiftly, that the *range of conditions used* can have important and disastrous effects upon the results obtained.

This question of *range effect* has been investigated by my ex-colleague Dr E. C. Poulton; and I cannot do better than to quote his considered judgment, given in a personal communication to me: "When each subject receives a number of treatments in a balanced or random order, unwanted range effects can sometimes *reverse* the rank order of the experimental results. In a *response* range effect, individual responses are influenced by the range of responses. In a *stimulus-response* range effect, the range of responses are matched by the range of stimuli. In a *stimulus* range effect, the stimulus is influenced by the range of stimuli. Range effects generally involve a central tendency [i.e., a tendency for results to cluster around the median of the total range] but not always. There is no way of discovering whether a within-subject design has introduced an unwanted range effect, except by repeating parts of the experiment using a separate-groups design."

This is a serious and sweeping indictment indeed; but unfortunately, as we shall see, it is well founded. Poulton cites very many examples to support his contention; we shall only consider a couple.

An experiment was conducted (Kennedy and Landesman, 1963) to determine the optimum height of a work bench for a standing operative. Two studies were conducted, both involving a "balanced" design with sixteen subjects working at six different heights of table (for each individual subject the six heights were at the same set of distances above and below the natural position of his elbow). In one study (A), however, the six heights were from 10 in. below elbow height to 10 in. above, at 4 in. intervals, in the other (B) the six heights were from 18 in. below to 2 in. above. The results were enlightening: study A showed that the optimum height was between the (statistically distinguishable) +2 in. and −2 in. positions; but study B indicated an unambiguous optimum at 6 in. below elbow height. Thus, while study A showed that a height of 2 in. above elbow height was clearly superior to one 6 in. below, study B reversed this order.

This result might be described as a simple central tendency resulting in a reversal of rank order. The second example we shall look at also reverses a rank order; but the range effect is asymmetrical.

In a simple reaction time experiment, subjects have to respond (e.g., by pressing a button) whenever a stimulus (e.g., a light) appears. There are a number of studies, however, which are slightly more complex than this. In these, there is a warning, or alerting signal (e.g., a white light) which precedes the action stimulus (e.g., a red light) to which the subject must respond. The time between the warning and action signals is called the foreperiod; and characteristically the reaction times are notably shorter in such experiments than in the "simple" ones.

If separate groups of subjects are tested with different durations of foreperiod, it is found that reaction time drops sharply as the foreperiod is increased from 0 to 0.25 sec. but thereafter it gradually, but reliably, increases again. However, this finding does *not* hold when each subject is presented with a variety of foreperiods in a "balanced" design. Then the longer foreperiods give reliably shorter reaction times than do the shorter; also, the differences are smaller.

We can intuitively understand this finding after a little thought. It is reasonable to imagine that, when presented with a varying series of foreperiods, a subject forms a rough estimate of an average value, and comes to expect it. When the foreperiod is notably shorter than average, he is likely to be unprepared, and to react somewhat more slowly, but when it is longer he is ready and waiting. Thus we find a reversal of rank order due to an asymmetric range effect.

These two examples should serve as grizzly warnings, especially in the light of Poulton's remarks that in surveying "virtually the whole of human experimental psychology [it is found] that range effects are a general characteristic of a man serving in an experiment with a within-range subject design." Why then, the reader may well ask, do we study this sort of design, and its associated techniques at all? There are two reasons: one base, the other somewhat less so.

The base reason is, to put it quite crudely, that examiners still demand it, and have not yet (1974) got round to noticing how silly it is. The less base reason turns on the limits within which experimenters in the human sciences have to work. As I remarked before in this book, it is true of research as of war that you do not do it as you would like to, but as you must.

Suppose that you have to investigate the effect of varying the size of some instrument dials under varying intensities of light: let us imagine that we have to consider six values of each. There would therefore be thirty-six possible combinations of size and lighting. It is hardly feasible to make reasonably valid comparisons of groups of less than six individuals, at the very least; so that a thorough investigation using separate groups would require a very minimum of 216 subjects — without making any allowance for individuals who run out, are run over, or run in. If the reader has ever tried to organize the arrival and departure within reasonably close time limits of 200—250 volunteers, he will readily understand why experimenters shrink from the prospect.

The seductive ease of a Latin Square design is doubly attractive in contrast: six subjects only, with one or two to spare for contingencies, are all that would be required. Nevertheless, the reader may persist: what avail is this economy in subjects if the result be fallacious?

Any answer necessarily smacks of special pleading; but is not to be dismissed for that. We cannot, indeed, rely on our rank ordering of the values of a particular variable if we are after an optimum value; but perhaps we can compare the relative importance of two variables. In our imaginary example, for instance, we might find that size of dial face accounted for notably more of the total variability than did light intensity — and, even bearing in mind the lurking possibility of interaction terms, this is no light thing.

Let us suppose that we accepted this tentative conclusion, it would then be reasonable to run (say) five to six separate groups with different values of *one* variable, in order to find the optimum value. In this way, the minimum subject requirement might be 6 (for the original "Balanced" design) + 36 (for the optimization) = 42, plus a few for contingencies. This is still a great saving on 216 + contingencies.

Thus, it must not be forgotten that, though we must regard the rank-

ordering of conditions with a sceptical reserve, analysis of Variance can give
a pointer towards the relative importance of various factors: for example
(in the example we described) of inter-subject differences of presentation
upon performance at some task. I say "a pointer towards" rather than a
"measure of" deliberately; the reader must remember the questionable
assumption of independence of factors.

The medieval Church used to refer to some revered saints as "authori-
ties of risk". This meant, when stripped of a lot of high-sounding Latin,
that no one would condemn you for citing them, but that their support
would not save you from burning. By analogy, we might call Analysis of
Variance a "technique of risk"; for while no editor will object to your
using it, it will not necessarily save you from awful howlers.

It will scarcely have escaped the readers' attention that Dr Poulton's
observations on the pitfalls of "balanced" designs apply to *any* balanced
design. It was for this reason that we have throughout this book repeatedly
expressed reservations about designs intended to "balance" for order of
presentation. The reader will now appreciate these cautions. Analysis of
Variance is, however, peculiarly liable to be disturbed by these range
effects, since an attempt is made to balance so many factors at once.

In this chapter we have examined certain experimental designs, namely
randomized blocks and *Latin Squares*, which aim to balance out, or
eliminate the possible unwanted order and transfer effects which can mar
investigations. We have seen that, though they cannot fully achieve this
end, they can be useful as general pointers and labour-savers. The Latin
Square design in particular aims to examine the relative importance of
individual differences and one other variable in the performance of a class
of tasks.

When using these designs we may employ Analysis of Variance. To do
this we first compute the *Total sum of squares* using (11.1), then the sums
of squares due to each of the other sources of variation being studied,
using (11.2) and (11.3), and also, if necessary, an exactly analogous expression for a further source, if one is being examined, just as shown in
(11.5). We then find the residual sum of squares, given (11.4), and which
is the difference between the total sum of squares and the sums of squares
associated with the several contributing factors.

We now find the variance estimate for each sum of squares, which is the
relevant sum of squares divided by the associated number of d.f.

The *F*-test is then used to determine the significance of the contribution
for each factor, *F* being the ratio:

$$\frac{\text{Variance estimate of factor}}{\text{Variance estimate of residual}}$$

F may also be used to compare substantial sources of variability, one with another.

Exercises

The questions to this chapter are less tests of your wits than to see whether you have mastered the necessary drill (sorry).

1. Two tests, each supposed to measure verbal facility, have been administered to six subjects. Three subjects had test A followed by test B, the others B followed by A. The scores obtained are given in Table 11.E1. Do they suggest that there are important differences between tests, or between subjects, or both? If both, which is more important?

Table 11.E1   Scores obtained by six subjects in two verbal fluency tests

	Subject					
	1	2	3	4	5	6
Score in test A	68	90	43	56	62	80
Score in test B	55	89	43	47	60	63

2. There is some evidence that subjects' ability to perform monitoring (vigilance) tests varies with time of day. Four subjects have each been given four one-hour vigilance tasks (A–D) at four different times (I–IV) in Latin Square order. Table 11.E2 gives the number of omissions registered by each subject during each test. Is there evidence that time of day is (*a*) an important variable, (*b*) the most important variable?

Table 11.E2   Latin Square design, and results of a vigilance experiment (scores are numbers of omissions)

		Time			
		I	II	III	IV
Subject	1	A9	D11	C13	B17
	2	C2	A4	B7	D8
	3	B26	C20	D22	A33
	4	D10	B22	A20	C28

3. An investigator is interested in the effect of background noise on performance in a test of manual dexterity. As a pilot experiment he takes three subjects and gives them three tests of manual dexterity (the scores

being time taken to perform each task to a criterion) varying the order of presentation in Latin Square. Table 11.E3 gives the results in order I—III for the three subjects.

Table 11.E3   Latin Square design and results of an experiment on manual dexterity (scores are time in seconds)

		Order of test		
		I	II	III
Subject	1	A 280	B 280	C 300
	2	B 315	C 290	A 283
	3	C 300	A 270	B 310

Comment on these results.

References

Kennedy, J. E. and Landesman, J. "Series effects in motor performance studies", *J. Appl. Psychol.*, 1963, **47**, 202—5.

Poulton, E.C. "Unwanted range effects from using within-subject experimental designs", *Psychol. Bull.*, 1973, **80**, 113—21.

12

Tendency and agreement

In all the tests we have examined so far, the kind of questions we have been asking have been essentially concerned with *relative magnitude*. We have been asking: is A greater than B? is effect X negligible or not? is the contribution of P to some outcome as important as that of Q? These are important questions indeed; but there are other classes of question we would like to answer.

These other sorts of questions may generally be called questions of *association*. They ask: does A go with B? They are questions of Tendency and Agreement: is it the case that heavy smoking goes with ill health? that big people have mild dispositions? that there is substantial agreement among persons who have read them that some given set of novels have some stated rank-order of merit?

None of the techniques we have so far studied would enable us satisfactorily to answer these questions: we need others.

Let us suppose that a set of ten subjects has been ranked for two qualities A and B. Ten others have been ranked for properties X and Y. In either case we can construct a diagram in which a subject's rank in each quality are taken as points on an axis, so that each subject has a point in the *xy* plane determined by his two ranks. For these two quite imaginary cases this has been done in Fig. 12.1 (*a*) and (*b*).

It does not require more than a glance at the figures to see that a subject's A-rank is singularly uninformative about his B-rank, while, on the other hand X-rank clearly tells us a lot about Y-rank. In technical usage, X and Y are *highly correlated*, while A and B are not. We need a measure of CORRELATION.

We are now going on to consider two measures of correlation: Spearman's ρ (Greek letter rho) and Kendall's τ (Greek letter tau). Before we do so, let us consider the general characteristics we require of such a measure.

If it were the case that two qualities had identical ranks — i.e., the subject who had some rank for quality P always had the same rank for quality Q — we could describe the two sets of ranks as perfectly correlated.

Fig. 12.1   (a) A/B positions of 10 Ss
(b) X/Y positions of 10 Ss

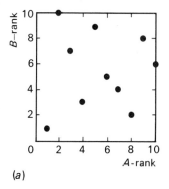

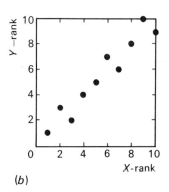

(a)   (b)

We may require our correlation measure to yield a value 1 for this case. Suppose, however, that the two qualities were such that their rank order were completely reversed. In this case, out of n subjects, the man with rank 1 for A would have rank n for B, the man with rank 2 for A would have rank $n - 1$ for B, and so on to the man with rank n for A and 1 for B. In this case, knowledge of A-rank also gives immediate and perfect knowledge of B-rank; and we call this case *perfect negative correlation*: our measure must have value -1. In between these cases, we must pass a point where there is no relation at all between A and B, and the distribution of points on a plane would be purely random. In this case, the measure must have value 0.

The mathematical model for measures of correlation is interesting and informative, and not very difficult. We shall therefore give it here; but the reader who is not concerned may take all on trust and skip from the beginning of the next paragraph (marked *) to that marked ** on p. 131 below.

* Suppose we have a set of n objects, which we are considering with respect to two properties X and Y. Let us number these objects $1 \ldots n$, simply to tag them: this has nothing to do with their ranks in either property. Let the rth object have value x_r in X, and y_r in Y. (These can be ranks or values of a variable; though we shall only consider ranks.)

Now let us consider any *pair* of objects, let us say the ith and the jth, and to that pair we allot an x-score denoted by a_{ij}. Note that this term is a property of the *pair*, and we make only one condition, namely that $a_{ij} = -a_{ji}$. (For example, we might be considering x to be mass in grams and a_{ij} to be the difference in mass between x_i and x_j. Alternatively, a_{ij} might be a difference in rank.)

Similarly, let the y-score of the pair be b_{ij} with the same restriction. Let us further state that if $i = j$, $a_{ij} = 0$.

We can now write down a generalized Coefficient (measure) of
Correlation:

$$C = \frac{\Sigma(a_{ij}\, b_{ij})}{\sqrt{[\Sigma(a_{ij}^2)\, \Sigma(b_{ij}^2)]}} \tag{12.1}$$

The reader can satisfy himself that this expression satisfies the general
criteria we discussed above. For example, if $a_{ij} = b_{ij}$ (perfect correlation)
then substituting in (12.1) gives $C = 1$, and so on.

Now let p_i and p_j be the *ranks* of the ith and jth object respectively in
quality X, and let q_i and q_j be their ranks in quality Y.

Also let

$$a_{ij} = p_i - p_j$$

and let

$$b_{ij} = q_i - q_j$$

Now p and q both range from 1 to n, so that all the terms $(p_i - p_j)$ are
formed by a permutation from the numbers $1 \ldots n$, two at a time, hence
the sums of squares

$$\Sigma(p_i - p_j)^2 \quad \text{and} \quad \Sigma(q_i - q_j)^2$$

are equal. Hence substituting in (12.1) gives:

$$C = \frac{\Sigma(p_i - p_j)(q_i - q_j)}{\Sigma(p_i - p_j)^2} \tag{12.2}$$

But:

$$\sum_{j,\, i=1}^{n} (p_i - p_j)(q_i - q_j)$$

$$= \sum_{i=1}^{n}\sum_{j=1}^{n} p_j q_j + \sum_{i=1}^{n}\sum_{j=1}^{n} p_j q_j - \sum_{i=1}^{n}\sum_{j=1}^{n} (p_i q_j + p_j q_i)$$

$$= 2n \sum_{i=1}^{n} p_i q_i - 2 \sum_{i=1}^{n} p_i \sum_{j=1}^{n} q_j$$

which (recalling that both $\sum_{i=1}^{n} p_i$ and $\sum_{j=1}^{n} q_j$ are merely the sum of the first
n natural numbers)

$$= 2n \sum_{i=1}^{n} p_i q_i - 2 \sum_{i=1}^{n} i \sum_{j=1}^{n} j$$

$$= 2n \sum_{i=1}^{n} p_i q_i - \frac{n^2}{2}(n+1)^2 \tag{12.3}$$

Let us define d as $(p_i - q_i)$ then

$$\Sigma d^2 = \sum_{i=1}^{n} (p_i - q_i)^2 = 2 \sum_{i=1}^{n} p_i^2 - 2 \sum_{i=1}^{n} p_i q_i \tag{12.4}$$

and therefore from (12.3):

$$\sum_{i,j=1}^{n} (p_i - p_j)(q_i - q_j) = 2 \sum_{i=1}^{n} p_i^2 - \frac{n^2}{2}(n+1)^2 - n \sum_{i=1}^{n} d^2 \tag{12.5}$$

But Σp_i^2 is the sum of the squares of the first n in natural numbers, i.e.,

$$\frac{n}{6}(n+1)(2n+1)$$

therefore

$$\sum_{i,j=1}^{n} (p_i - p_j)(q_i - q_j) = \frac{n^2}{6}(n^2 - 1) - n \Sigma d^2 \tag{12.6}$$

Also

$$\sum_{i,j=1}^{n} (p_i - p_j) = 2n \sum_{i=1}^{n} p_i^2 - 2 \sum_{i,j=1}^{n} p_i p_j$$

$$= 2n \sum_{i=1}^{n} p_i^2 - 2 \left(\sum_{i=1}^{n} p_i \right)^2$$

$$= \frac{n^2}{6}(n^2 - 1) \tag{12.7}$$

Substituting from (12.6) and (12.7) in (12.2):

$$C = 1 - \frac{6 \sum\limits_{i=1}^{n} d^2}{(n^3 - n)} \tag{12.8}$$

This expression defines the particular correlation coefficient we have mentioned as Spearman's ρ.

** To those who have just rejoined us, we can explain that we have proved that Spearman's ρ satisfies the criteria we set up for a coefficient of correlation. We have shown that if a population of n items is ranked for qualities x and y, and if the ith object has rank p_i in x and q_i in y we may take the difference of these ranks

$$d_i = (p_i - q_i)$$

and set up a correlation coefficient:

$$\rho = 1 - \frac{6\Sigma d^2}{(n^3 - n)} \tag{12.8}$$

We shall now give an example of the use of ρ.

The Serious Minded Cannibal we have met before found that some missionaries had given ten of their converts a scripture exam, whose marks he obtained. These converts were subsequently eaten by the SMC and his friends, each of whom independently awarded each carcass an edibility score. The SMC decided to take the mean of these edibility scores as each convert's true edibility value; and was interested to see whether or not edibility correlated positively with scriptural learning. The scores are given in Table 12.1.

Table 12.1   Scripture and edibility marks obtained by ten subjects

	Subject									
	1	2	3	4	5	6	7	8	9	10
Scripture mark	32	40	51	55	56	62	64	70	80	91
Edibility	27	11	55	50	41	30	80	75	95	68

He decides to employ ρ to investigate his problem; and, being in a cautious mood, sets an α of 0.02.

The procedure is simplicity itself. We construct a table in which the first column lists the subjects, the second each subject's *rank* for one quality (here scriptural knowledge), the third his *rank* for the other (edibility), the fourth the difference, d, between these ranks, and the fifth and last lists the several d^2 values. Simple addition gives us Σd^2; n is the number of subjects, and substitution in (12.8) gives us ρ. (See Table 12.2).

Here we see that $\Sigma d^2 = 36$; and hence

$$\rho = 1 - \frac{6 \cdot 36}{(1000 - 10)} = 1 - \frac{216}{990} = \frac{774}{990} \doteq +0.78$$

Consulting Appendix Table M, we find that, for $n = 10$, we can reject H_0 (that the correlation is not truly greater than zero) at the 0.01 level: the SMC's α is amply satisfied: there is a positive correlation between edibility and divine learning.

The reader will have noticed that Appendix Table M doesn't help if $n > 30$. We may well find ourselves, however, wishing to use much larger numbers; e.g., in social statistics we may find variables tabulated over fifty years or more. In such cases we may use the transformation:

$$t = \rho \sqrt{\left[\frac{n - 2}{1 - \rho^2}\right]} \tag{12.9}$$

Table 12.2   Data of Table 12.1 set out for the computation of

Subject	Rank of scripture mark	Rank of edibility score	d	d^2
1	1	2	−1	1
2	2	1	1	1
3	3	6	−3	9
4	4	5	−1	1
5	5	4	1	1
6	6	3	3	9
7	7	9	−2	4
8	8	8	0	0
9	9	10	−1	1
10	10	7	3	9
			$\Sigma d^2 = 36$	

t, here being strictly equivalent to the "Student's t" we looked at in Chapter 10, and we may therefore consult Appendix Table K for $p|H_0$).

Another problem which may arise is that of tied ranks. If these are few — say, one batch of two or three in one quality on an n of 10 or so — there is no cause to worry. If, however, they are numerous, a correction must be applied.

Suppose that, at some given x-rank, t values are tied. We form the quantity

$$T = \frac{t^3 - t}{12} \qquad (12.10)$$

The sum of all the Ts for the x-quality will be ΣT_x and similarly the sum of all those for the y-quality (which will, of course, in general be different) will be ΣT_y.

We now need form two new quantities, namely:

$$N(x) = \frac{n^3 - n}{12} - \Sigma T_x \quad \text{and} \quad N(y) = \frac{n^3 - n}{12} - \Sigma T_y$$

and then:

$$\rho = \frac{N(x) + N(y) - \Sigma d^2}{2\sqrt{[N(x) \cdot N(y)]}} \qquad (12.11)$$

The reader will readily agree that this computation may be a trifle tedious, but does not involve any real difficulty.

A very important caveat must be entered concerning the use of measures of correlation, such as ρ. When two quantities correlate highly together, there *may* be a causal link between them; but there may not. It may be that both quantities depend upon some third factor, or complex of factors, and do not directly affect one another at all. It may be happenstance. We can easily cite instances of all three; and equally easily, we can think of disputed cases.

For example; since volume is proportional to the cube of linear dimensions, it need surprise no man that height and weight are highly positively correlated (not perfectly: there are very lean tall people and gross short ones) but, to take a contrary case, there is no reason to suspect a causal relation because the sale of TV sets correlates rather highly with the number of persons in mental hospitals. The admittedly high correlation between consumption of cigarettes and lung cancer was for long a matter of dispute: was it causal connection, separate effects of some third factor (as, e.g., Professor Eysenck maintained) or even fortuitous? For technical reasons we need not discuss here, I think very few competent persons would now question that the connection is indeed causal.

The reader must be on his guard, therefore: for if he should find a very high positive correlation, this *may or may not* by itself indicate a causal relationship. Broadly speaking, we need some independent reason to suspect such a relationship before we can regard correlation as affording direct evidence.

A method of measuring correlation alternative to ρ is Kendall's τ. Its rationale is again of delightful simplicity; and we shall give it, again between one and two asterisks.

* In (12.1) above let the rank of the ith individual in x quality be p_i and that of the jth member be p_j. Then let us give the x-score a_{ij} one of two values:

if $p_i > p_j$, $a_{ij} = +1$

if $p_i < p_j$, $a_{ij} = -1$

and similarly for b_{ij}, the y-score. Therefore, if both $(p_i - p_j)$ *and* $(q_i - q_j)$ have the *same* algebraic sign $a_{ij} b_{ij} = +1$; but if they have *opposite* signs, $a_{ij} b_{ij} = -1$.

Let us take all the scores of different pairs and add them, obtaining a value S. We will find that $\Sigma\, a_{ij}\, b_{ij} = 2S$, since any given pair occurs twice in the summation: once as (ij) and once as (ji).

Also

$$\Sigma\, a_{ij}^2 = \Sigma\, b_{ij}^2 \quad [\text{for } (+1)^2 = (-1)^2 = 1]$$

therefore

$$\Sigma\, a_{ij}^2 = \Sigma\, b_{ij}^2 = n(n + 1)$$

Whence, substituting in (12.1):

$$\tau = \frac{2S}{n(n-1)} \tag{12.12}$$

** For those who were prepared to take our word for it, what we have proved is this: if n individuals are ranked for two qualities x and y, we may take all possible different pairs of individuals, and if the difference in their ranks for *both* qualities has the *same* algebraic sign we give that pair the score +1, and if the difference in ranks for the two qualities has *opposite* algebraic sign we give that pair the score -1, we can sum the scores over all pairs, $= S$, and arrive at a valid coefficient of correlation:

$$\tau = \frac{2S}{n(n-1)} \tag{12.13}$$

This might seem a little tedious at first, since the number of different pairs goes up steeply with n: it is in fact equal to $(n/2)(n-1)$. However, there are, fortunately, a number of nice short-cut ways of finding S without laboriously going through all possible pairs. We shall now describe one of the simplest.

First, set out the data so that one set of ranks, say the X-ranks, are in their natural order (1 on the left, to n on the right). Set out the corresponding Y-ranks. Take the first Y-rank, and count the number of Y-ranks to its right that are *larger*: subtract the number that are *smaller*: repeat this for each successive Y-rank, and the sum so obtained will be S.

This is very readily shown by an example. If we refer back to the computation carried out by the Serious Minded Cannibal, we can lay out the data of Table 12.2 somewhat differently.

Table 12.3   Data of Table 12.2 set out for computing τ

	Subject									
	1	2	3	4	5	6	7	8	9	10
Scripture rank	1	2	3	4	5	6	7	8	9	10
Edibility rank	2	1	6	5	4	3	9	8	10	7

Taking successive edibility ranks from the bottom of Table 12.3, we have:

$$S = (8 - 1) + (8 - 0) + (4 - 2) + (4 - 2) + (4 - 1) + (4 - 0)$$
$$+ (1 - 2) + (1 - 1) + (0 - 1)$$
$$= 7 + 8 + 1 + 2 + 3 + 4 - 1 + 0 - 1$$
$$= 23$$

therefore

$$\tau = \frac{46}{90} \doteq 0.51$$

The reader will immediately notice that numerically $\tau \neq \rho$. At the least, this means that the measures must not be confounded nor directly compared. Another important question is: will a particular outcome which, by one measure, accepts or rejects H_0 at a particular level similarly accept or reject it, at that level, by the other measure? Unfortunately, this is a somewhat moot point. It would seem that the desired equivalence is approached for large n; but it is not evident how large is "large".

At this point the reader will probably be asking: if ρ and τ are equally usable and about equally powerful, is there any reason for preferring one to the other in any given case? As a rule, the answer is "no": it is merely a matter of taste and convenience. There is, however, one objective point in favour of ρ. That is, that it bears a regular relation to the concordance coefficient W, which we are about to discuss, while τ does not. But if we are not going to, or do not need to use concordance measures, this is not germane to our decision. Personally, I always use ρ, with which I find I make fewer mistakes in arithmetic; but I do not advance that as more than a personal quirk.

As with ρ, there are corrections to τ which must be applied for ties, and there is an expression which enables us to use it with large n.

Suppose that, at some X-rank, t values are tied. We form the quantity $(t^3 - t)$, and, if there are more than one set of ties, the quantity $T_x = \frac{1}{2}\Sigma(t^3 - t)$. Similarly for the Y-ranks, we may form the quantity T_y. Then

$$\tau = \frac{S}{\sqrt{\left[\frac{n}{2}(n - 1) - T_x \quad \frac{n}{2}(n - 1) - T_y\right]}} \tag{12.14}$$

When $n > 10$, it can be shown (though the proof is outside our scope) that τ has a Gaussian distribution with

$$\mu = 0$$

$$\sigma = \frac{2(2n + 5)}{9n(n - 1)}$$

Whence it is possible to transform τ to a new variable z, given by:

$$z = \frac{\tau}{\sqrt{\left[\dfrac{4n + 10}{9n(n - 1)}\right]}} \qquad (12.15)$$

and to consult Appendix Table A for the associated probability.

Thus far we have considered an agreement (or otherwise) between *two* sets or rankings. However, we often need to examine more than two such sets. For example, suppose six candidates are interviewed by five persons who have to award a job to one of them. In general, they will not agree exactly among themselves; but it would be useful to have a measure of their agreement, and to be able to see how probable such a degree of agreement is. For such problems we use Kendall's coefficient of concordance W. Let us suppose that our five judges have independently ranked our six candidates, as shown in Table 12.4.

Table 12.4   Rankings of six candidates by five independent judges

	Candidate					
	a	b	c	d	e	f
Judge A	1	2	3	4	5	6
B	2	1	4	3	6	5
C	1	3	2	4	6	5
D	2	5	1	3	4	6
E	1	4	3	2	6	5
R_i	7	15	13	16	27	27

At this stage it is worth sitting back once again and considering the general characteristics our coefficient must have.

One point should be noted immediately: as soon as there are more than two sets of rankings, there cannot be *total* disagreement. Although Judge A may give a candidate rank 1 and Judge B, in total disagreement, can give him rank n, Judge C cannot totally disagree with *both*; he can select any value from 1 to n; but he cannot flatly contradict both his colleagues, however much he may want to. Thus, although the several rankings may be wholly random, they cannot be less than that. In other words, we expect our coefficient to range from 0 to +1 and *not* from −1 to +1.

Further, if we consider a table such as Table 12.4 giving the rankings by n judges of m candidates, we can state what we would find under certain conditions. If all judges were in perfect agreement, the first candidate would always be ranked 1, and the first column total R_1 would be n; the second would be $2n$, and so on to the last, which would be nm. On the other hand, if there were no agreement, we would expect the column totals to be about equal.

Let us pursue this a little further. Each row of the table will consist of the first m natural numbers − however ordered − and must therefore sum to $\frac{1}{2}m(m + 1)$. Therefore, the sum of all entries, which is, of course, the sum of all column totals

$$= \sum_{i=1}^{m} R_i = \frac{nm}{2} (m + 1)$$

And, since there are m columns, the mean column total

$$= \bar{R} = \frac{n}{2} (m + 1)$$

As remarked above, if all judges agreed, the several R_i would be n, $2n \ldots nm$ and hence their deviations from the mean would be:

$$n - \frac{n(m + 1)}{2}, \quad 2n - \frac{n(m + 1)}{2}, \ldots, nm - \frac{n(m + 1)}{2}$$

which become:

$$-\frac{n}{2} (m - 1), \quad -\frac{n}{2} (m - 3), \ldots, +\frac{n}{2} (m - 1)$$

and the sum of the squares of these deviations is:

$$\frac{n^2}{12} (m^3 - m)$$

This is the *largest* value which the sum of the squares of the deviations of the column totals *can* have. Evidently, if judges responses are merely random, the column totals will all be equal, and the sum of the squares of the deviations will be zero. Now let the actual sum of squares of deviations in a particular case be S, i.e.:

$$S = \Sigma(R_i - \bar{R})^2 \tag{12.16}$$

We may now define our COEFFICIENT OF CONCORDANCE, W, as the ratio of this quantity to the maximum value possible, which we have already found, i.e.,

$$W = \frac{12S}{n^2(m^3 - m)} \tag{12.17}$$

We can now return to the values in Table 12.4. Here $n = 5$, $m = 6$ and $\Sigma R_i = 105$; whence $\bar{R} = 17.5$.

The several values of $(R_i - \bar{R})$ are -10.5, -2.5, -4.5, 1.5, 9.5 and 9.5, hence:

$$S = 319.5$$

Substituting in (12.17) gives

$$W = \frac{12 \cdot 319.5}{5^2(6^3 - 6)} = \frac{639}{875} = 0.73$$

Our H_0 is evidently that the several rankings have only random agreement, one with another. We may consult Appendix Table O and find that

$$p \mid H_0) < 0.01$$

There are a number of points to consider here. First, when we have found a W and rejected our H_0 at whatever our chosen α may have been, what are we to make of the final result? There is some force in the rhetorical question: "So the judges are in substantial agreement: so what?" If we suppose that the judges were reasonably competent, or a reasonably representative sample of some larger population whose opinions we wished to know, what is the best estimate we can make of the "true" ranking, whatever that may be?

It does not seem possible to improve on Kendall's suggestion that the best procedure is to rank according to the sums of the ranks allotted to each individual — i.e. (as we have laid out our data in our table), according to column totals. Thus, we may rank our candidates in order as a, c, b, d and e = f.

Appendix Table O, as the reader will have noted, does not take us beyond an m of 7. For $m > 7$, it is necessary to form the quantity X, given by

$$X = n(m - 1)W \tag{12.18}$$

This can be treated as χ^2 with $(m - 1)$ d.f., and Appendix Table D may be consulted.

It sometimes happens that a judge is unable to distinguish between two or more individuals, producing, in his row of the table, a set of tied ranks. If the number of such ties is small, they may safely be neglected; but if they are numerous a correction should be applied. (Unfortunately, I am unable to give the reader a set number or proportion which should be treated as "numerous": it is necessary to use commonsense: personally I regard more than one-third of at least one row as rough rule of thumb.)

Suppose that for any one Judge J, there are t ties at some given rank. If more than one rank has ties, we have, for that judge:

$$T_j = \frac{\Sigma(t^3 - t)}{12}$$

Computing this term for all judges, let

$$T = \sum_{j=1}^{n} T_j$$

We then modify the expression for the coefficient of concordance to:

$$W = \frac{S}{\left[\dfrac{n}{12} (m^3 - m) - nT \right]} \tag{12.19}$$

In discussing concordance measures we have talked consistently about "Judges" and "individuals". This is not, of course, a necessary limitation of the method. The reader should be clear that we could equally have talked of "experimenters" and "scores", or "estimators" and "objects". It is only necessary that some set of objects or persons should be independently ranked for some measure or quality by another set of independent persons or processes.

Let us imagine an investigation which makes use of both concordance and correlation measures. Let us suppose that there is a long-standing dispute among cultural historians (who do indeed dispute such things) as to whether or not the quality of artistic productions within society Z does, or does not, tend to rise and fall in rough step with the total production of wealth within that society.

The reader will immediately observe certain difficulties in the way of resolving such a dispute. The trouble is that, although one factor in the dispute – the GNP – is usually measurable, the other, at present, is not. How, then, can the problem be approached?

A possible (not, admittedly, wholly unquestionable) procedure would be to find a group of art experts not party to the argument, and ask each of them independently to rank the overall quality of the local equivalent of the Royal Academy summer exhibitions at (say) five-year intervals over (say) fifty years. If their opinions yielded a very high concordance, it would then be possible to see whether the overall ranking so found correlated with that for GNP for the same years.

Table 12.5 gives the rankings of six independent art experts of the exhibitions at years arbitrarily started at 0. (Lowest rank = highest quality.)

Table 12.5   Independent rank-orderings of aesthetic merit by six judges

	Year										
	0	5	10	15	20	25	30	35	40	45	50
Judge 1	1	3	5	7	8	9	10	11	6	4	2
2	1	2	4	5	7	6	8	10	11	9	3
3	2	5	6	8	9	7	10	11	4	3	1
4	3	5	6	7	9	8	11	10	4	1	2
5	1	2	3	7	8	9	11	10	6	5	4
6	3	5	8	9	10	11	6	7	4	2	1
R_i	11	22	32	43	51	50	56	59	35	24	13

Here, $n = 6$; $m = 11$; and $\bar{R} = (n/2)(m + 1) = 36$.
Using (12.16) we find that $S = 2910$; whence, substituting in (12.17):

$$W = \frac{12 \cdot 2910}{6^2(11^3 - 11)} \doteq 0.735$$

Since $m > 7$ we must use the equivalent given by (12.18) which yields us a value of

$$\chi^2 = 60W = 44.1$$

with $(11 - 1) = 10$ d.f.

Before consulting Appendix Table D, we should set our α. As there is a certain dubiousness about the whole business, perhaps we should set a rather strict α, and only reject H_0 if $p \mid H_0) < 0.01$. Consulting the table, we find that our α is satisfied; and we are justified, at least as a tentative measure, in using the R_i values to rank the several years for their artistic merit.

We can easily obtain values of GNP from regularly published works of reference (known in that society as the "Jahrbuch des statistics") and ranking these is a simple matter. Having rank orders for GNP and artistic merit we can now inquire whether they correlate. For this we may choose to use ρ, and lay out our rankings by year as shown in Table 12.6.

Table 12.6   Ranks for aesthetic merit and GNP set out for ρ

Year	Aesthetic rank	GNP rank	d	d^2
0	1	11	10	100
5	3	10	7	49
10	5	9	4	16
15	7	8	−1	1
20	9	6	−3	9
25	8	7	−1	1
30	10	5	−5	25
35	11	4	−7	49
40	6	3	−3	9
45	4	2	−2	4
50	2	1	−1	1

$$\Sigma\, d^2 = 264$$

Since $\Sigma\, d^2 = 264$ and $n = 11$, substitution in (12.8) gives:

$$\rho = 1 - \frac{6.264}{11^3 - 11} = 1 - \frac{1580}{1320} = 1 - 1.20 = -0.20$$

For the same reasons as those given above, we should probably select a fairly strict α, say 0.01. Alas, Appendix Table M shows that even on a very slack criterion we could not be justified in rejecting H_0: there is no reason to suppose that the two rankings are correlated at all. No man should be surprised at this.

In this chapter we have described methods of determining whether two or more sets of rank-orderings tend to be associated — i.e., measures of *correlation* and of *concordance*. If we can rank order two qualities for a set of individuals (which may or may not be human subjects: they may be times, societies, or anything to which the rankable qualities may be attributed) we have two measures for their correlation.

We may use Spearman's ρ which is given by (12.8). If the number of individuals is >20, the expression may be modified to (12.10) and probability examined by regarding this as a value of t. If there happens to be a lot of tied ranks, it is desirable to use the corrected expression (12.11).

Alternatively, we may use Kendall's τ which is given by (12.13) (a short-cut method of computation is given in the paragraph following the expression). If the number of individuals is > 10, τ may be transformed to the Gaussian variable z, given by (12.15); and when the number of ties is large the modified expression (12.14) should be used.

When we wish to examine the concordance of more than two sets of rankings, Kendall's coefficient W is recommended. We set out the n sets of m ranks, and find the rank total R_i assigned to each of the m individuals. We then form the term $S = \Sigma(R_i - \bar{R})^2$ and substitute in (12.17) for W. Should m be > 7, we form X using (12.18), which is treated as a value of χ^2 with $(m - 1)$ d.f. If tied ranks are numerous, W is modified as in (12.19).

In all questions where correlation and concordance are used, care must be taken to ensure that chance agreements are not accidentally inflated to an appearance of great consequence. As in all statistical problems, the use of formulas is no substitute for thought and judgment, though it can greatly aid them.

Exercises

1. A resourceful politician draws attention to a clear connection between the numbers of purple immigrants and the number of convictions for keeping a disorderly house. He supports his contention as in Table 12.E1. Comment on his assertions.

Table 12.E1 Politician's data

	Year							
	1965	1966	1967	1968	1969	1970	1971	1972
No. of purple immigrants (1000)	305	314	362	388	401	411	414	416
No. of convictions for brothel-keeping	711	753	746	802	915	922	920	927

2. In an experiment recently concluded, I obtained from a set of subjects estimates of probability (expressed as per cent) and of their confidence in that estimate, expressed on a 7-point scale from 1 (= quite certain) to 7 (= just guessing). I noted the *errors* which subjects made in their estimate; and tabulated them together with their confidence rating. Have you any comment on these findings?

Table 12.E2   Data from an experiment by M.H.

	Subject								
	1	2	3	4	5	6	7	8	9
Confidence rating	1	5	2	3	1	4	1	3	2
Magnitude of error (per cent)	44	2	37	32	42	27	47	30	37

3. A set of fifteen schoolchildren are examined in maths, physics, French, history, English and biology. Their marks are given in Table 12.E3. It is suggested that quality of performance is general rather than specific: i.e., that the good and bad in one subject will be good and bad in any other. Do the data support this view?

Table 12.E3   Exam results of a group of schoolchildren

	Child														
Score in	1	2	3	4	5	6	7	8	9	10	11	12	13	14	15
Physics	85	80	78	75	70	57	60	65	61	59	43	40	35	25	7
Maths	99	92	84	80	78	71	63	62	58	52	42	35	29	23	5
French	90	96	70	78	85	50	57	45	66	60	32	30	37	36	38
History	77	71	72	74	83	55	50	68	65	61	37	42	10	32	16
English	74	70	78	64	58	51	54	49	36	47	15	20	10	6	40
Biology	68	75	66	64	70	58	55	40	50	60	26	38	30	23	25

4. Six judges are asked to rank eight paintings for aesthetic merit. Their rankings are shown in Table 12.E4. Does any substantial harmony or disharmony of views emerge?

Table 12.E4   Rankings of eight paintings according to six judges

	Painting							
	a	b	c	d	e	f	g	h
Judge 1	3	1	4	5	2	6	8	7
2	6	8	4	5	7	3	2	1
3	4	1	3	5	2	6	7	8
4	3	2	4	6	1	5	8	7
5	6	7	4	5	8	3	1	2
6	5	8	6	4	7	3	2	1

Reference

Kendall, M. G. *Rank correlation methods*, London, Griffin, 1948.

Answers to exercises

Chapter 2

1. It is remotely improbable that necrophylia was more common then than now; and, in any case, that has nothing to do with it. The two populations whose means are given are different. The "mean age of death" is the mean age, at death, of *all* persons born alive. The "mean age of marriage" is the mean age of only those persons who both live until they can, and then in fact do marry. Since, in those happy times before industrial pollution and "anomie" had (if you care to believe it) started killing us off early, a handsome proportion of all those born alive died before they were five years old; this easily brought the mean age at death below the mean age of marriage.

2. Appendix Table A gives the 99 per cent as 2.33 s.d. above the mean. Thus 99 per cent of the male population will be less tall than $(173 + 2.33 \times 6.5)$ cm, $= 188$ cm. However, heights are generally measured in bare feet, with the measuring bar pressed onto the scalp. How much should you allow for shoes and hair? (My guess is 2 cm.) If you overlooked this point, lose five marks for being in an Ivory Tower: statistics is an *applied art*.

3. No, *not* the mean. True, the *population* I.Q. is very nearly Gaussian but the *tail* of that distribution is nothing of the kind (see Fig. 2.4). Even allowing for imperfections in the cut-off techniques, the mode will be very nearly at the lower limit. Therefore the *median* is the quantity to use.

Our estimate of the median can only be rough, because the distribution is not exactly that of the population tail (as will be obvious after a moment's thought). But, assuming for the argument that it is, we want the 96 per cent point (the cut-off for half of 8 per cent). Appendix Table A gives this as $\mu + 1.75\sigma$, which in this case gives an I.Q. of just over 126.

4. The mean of X-group's Q-scores is 2.4 s.d.s above the "threshold" for observing C — whatever that is. That of the Y-group's Q-scores is some 1.07 s.d. above it. Thus, consulting Appendix Table A again, we expect

approximately (100−99.2)% = 0.08% of all Xs to exhibit C, while some 14.2% of all Ys should do so. We thus expect that, in proportion to their absolute numbers, almost eighteen times as many Ys as Xs will have characteristic C.

As we remarked, truly Gaussian distributions are not too common in the Human Sciences; so in any real situation such a sum would be an approximation. We might conclude, however, that ratios markedly different from eighteen to one could not be accounted for by genetic factors.

Chapter 3

1. The third law of probability states that

$$p(B) = p(B \mid A)p(A) + p(B \mid \bar{A})p(\bar{A})$$

Here let B = "Survivor", and let A = "deals three court cards".

It is given that

$$p(B \mid A) = 0.2$$
and $p(B \mid \bar{A}) = 0.01$

$p(A)$ is easily computed. There are sixteen court cards in the pack, and hence the number of hands of three court cards is

$$^{16}C_3 = \frac{16!}{3! \ 13!}$$

The total number of all three-card hands is

$$^{52}C_3 = \frac{52!}{3! \ 49!}$$

therefore

$$p(A) = \frac{^{16}C_3}{^{52}C_3} \doteqdot 0.025$$

therefore

$$p(\bar{A}) = 1 - p(A) \doteqdot 0.975$$

therefore

$$p(B) = (0.20 \times 0.025) + (0.01 \times 0.975) = 0.015$$

Which is a pretty thin chance.

2. (*a*) This is a problem for Bayes' Theorem. Here H is the proposition "he dealt three court cards", and D is the datum "he survived". We have, from above, that $p(D) = 0.015$, that $p(H) = 0.025$ and that $p(D \mid H) = 0.20$.

Whence

$$p(H \mid D) = \frac{0.20 \times 0.023}{0.015} = \tfrac{1}{3}$$

(*b*) Ask him, of course. Or see if he is wet from swimming.

3. Surprisingly high. Birthdates are not mutually incompatible events; so we must use the second law to compute the probability. The probability that the first two children shall have *different* birthdays is clearly $\frac{364}{365}$. Given this, the probability that the third differs from the first two (i.e., p(Birthday A and Birthday B and Birthday C are different)) is $\frac{364}{365} \cdot \frac{363}{365}$. Proceeding until we get to the twenty-fourth child, we have the required probability of all birthdays being different is

$$\frac{364}{365} \cdot \frac{363}{365} \cdot \frac{362}{365} \cdots \frac{343}{365} \cdot \frac{342}{365}$$

This reduces to $\frac{23}{50}$. Thus, the probability that one birthday shall be the *same* for two children is $1 - \frac{23}{50} = \frac{27}{50} = 0.54$ or somewhat better than $\tfrac{1}{2}$.

4. Here $p(H_0) = 0.999$; $p(D \mid H) = 0.99$ and $p(D) = (0.1 \times 0.001 + 0.99 \times 0.999)$. Whence $p(H \mid D) = \frac{0.98901}{0.98911}$.

This is very little less than 1. I would stop worrying.

Chapter 4

1. The value of $\bar{x}$, as already computed, is 31.6. s, however, is not $\sqrt{(28)}$ but $\sqrt{\frac{28.31}{30}} \doteq 5.38$. Thus the best estimate of μ is 31.6, and there is a 95 per cent probability that μ lies between 20.8 and 42.4. Similarly, the best estimator for σ is 5.38, and there is a 95 per cent probability that σ lies between 4.01 and 6.75.

2. The statement is very suspect indeed. To list only a few points: (i) The readership of a newspaper is far from a random sample of the population, being highly selective for attitudes, social group, etc. (ii) The respondents to a poll are selective even within that group, being the ones interested enough to answer. (iii) The form in which the question is put has great influence upon the way people answer. The reader will readily think of other objections.

3. There is no one hard and fast answer to this question. We have to select thirty children out of (roughly) 900. Clearly, it will not do simply to pick a class in one school and test all the children therein. Schools differ in their catchment areas, staff and a host of factors which *may* (we cannot say definitely) affect the result. It is important, therefore, that the sampling procedure shall be random over all the children, which may be accomplished by taking three children from each school. The problem of which three is a separate one. It should never be left to teachers to select —

they can be relied upon to select non-randomly. A possible method would be to write the names of the thirty or so target children in each school on identical slips of paper, shake them up in a hat, and draw three. The reader may well hit on a better method.

4. Well, it's for you to judge; but a *powerful* test is one which avoids Type II errors. In the present case, that is one which minimizes the likelihood of a 50 dinars fine. Now every man must make his own decision, but I'd far sooner pick the fine than the poisonous reptiles; and consequently the *power* of the test would not be a principal consideration for me. I consider Isaac to be unwise.

Chapter 5

1. This is a straightforward question: is treatment B *better than* treatment A. The obvious measure is frequency of recovery; and the problem is manifestly one-tailed. Since the pay-off is serious, a very small α seems called for. ($p < 0.01$? The reader will have his own ideas.) An important precaution in all such cases: the test must be "double-blind", i.e., neither the patient, nor the person who administers the treatment, must know whether A or B is being used. All sorts of troubles arise if this precaution is neglected.

2. This is a tricky one; indeed I am not at all sure that there is a satisfactory way of going about it. Measurement is no problem: number of sets and frequency of crimes (or indictable offences) are suitable — or number of sets held without licence, if you wish to be ingenious. The difficulties — which may be insuperable — lie in isolating the variable. While the number of sets sold has increased, a host of other factors which *may* be as, or more, important has changed also: standard of living, distribution of population, and the rise of "age group cultures" to name but three. While it were possible to compare changes in parts of the country which received TV coverage later than others, there is the objection that these areas were different in other ways: e.g., they were mainly rural. The reader may have good ideas of his own; and I shall only say that I am highly dissatisfied by all studies of the social effects of TV that I have read.

3. Here the problem is to see that the probability of the lower tail of the purity distribution cut off by the 96 per cent bound is sufficiently negligible. "Negligible" here defines the α required, and will depend on the pecuniary loss, and that in "good will", which would follow the despatch of a defective item. From this α, a tolerable σ will be derived; and hence, from consideration given in Chapter 4, an acceptable s for a sample of size n. The statistician should therefore select a random sample of size n from each batch before it goes to the wholesalers, and check the value of s: a routine, but important activity.

Chapter 6

1. In this case, if there were no difference between the accident-proneness of male and female drivers, we would not expect equal numbers, simply because an unequal number were at risk. Since 60 per cent of drivers were men, we would indeed suppose that, for a given accident p(man) = 0.6 and p(woman) = q = 0.4. We are asking, quite evidently, is p > 0.6? — i.e., is the probability of an accident happening to a man greater than the numerical disparity would suggest. In other words, this is a one-tailed question. α is, as ever, a matter of choice: I suggest 0.05 for the time being. n = 51 (>25) and we must therefore use a Gaussian approximation. Fortunately, npq = 51 . 0.6 . 0.4 = 12.24 which is >9, and the approximation can be used. Here x = 41, and substituting in (6.7):

$$z = \frac{41 - 51 . 0.60 - 0.5 .}{(12.24)} = 0.8$$

Whence Appendix Table A gives the probability of a number as high as 41 as $(1 - 0.79) = 0.21$. It would seem, then, that there is indeed no strong evidence for the greater accident-proneness of male drivers.

However, it will not have escaped the reader's attention that other questions can still be asked; e.g., although men and women were balanced in space (by selecting a particular stretch of road) were they so in time? i.e., did women tend to appear when traffic overall was less dense? Life is not easy for the human scientist.

2. This is obviously a two-tailed question, with no *a priori* weightings. The decorator merely wishes to know is p = q? As no great moment seems to attach to the question, he may well be satisfied with an α of 0.1. If we call an instance where the A-score > B-score "+", A < B "−" and A = B "=", we note from the table that, in an n of 22, we have 2 +s, 14 −s and 6 =s. The last group reduces the effective n to 16; and we now consult Appendix Table C for the probability of 2 or fewer +s.

The entry is 0.002; but this, we recall, is for the one-tailed question, so our value must be 0.004. This is well under the rather easy α; and the decorator may be advised to select B.

3. This really won't do as an experiment at all. Although the p associated with one disagreement out of twenty is <0.001, there are too many sources of doubt. Did the SMC balance for the order of presentation of the curries? Did his own judgment influence that of his guests? Did they know which was which? Were they somewhat apprehensive of disagreeing with their anthropophagous host? Until these questions are satisfactorily answered we must suspend judgment and there is no point in doing sums.

4. This is a two-tailed question, with n reduced effectively to 9 by the draws. Appendix Table C shows that the two-tailed probability of as few

as two wins by either player is 2 x 0.062 = 0.124. It depends, then, on the
α set; but even with an easy α of 0.1, it seems at first that the enthusiast
can reasonably stick to his guns. However, there is an important *caveat* to
be entered: especially towards the end of a match, the result of a particular game is by no means independent of the previous results. Thus the
validity of the binomial test in this instance is highly suspect, and it should
not be used.

Chapter 7

1. The table prepared by the learned doctor seems at first well-suited to
the application of χ^2; but the reader will have seen that two of the eight
cells have E values < 5. However, a perfectly reasonable "pooling" is
possible, if we divide the drinks into "alcoholic" and "non-alcoholic". The
resulting 2 x 2 table (with naturally, 1 d.f.) has a smallest E of 14.2. It is
difficult to set α: the hypothesis that the drinks are relevant seems *a priori*
reasonable; so a value of 0.05 may be acceptable. (7.4) can be used to
compute χ^2, which is approximately 44.4, indicating that $p(H_0) < 0.001$.
The learned doctor's *Speculations* seem'd Reafonable.

2. The point which catches the eye is the apparently marked change in
the relative frequency of fighter and bomber aircraft as the war continued.
This is well-attested from a variety of information, and we might well
think an α of 0.1 adequate. χ^2 with a d.f. of 12 is straightforward if tedious
to compute; and comes out as 28.4. Not surprisingly, this gives
$p(H_0) \ll 0.01$. The Historians seem to be right.

3. Evidently $m_1 = 14, m_2 = 11, r = 21$; so that $p(H_0) < 0.05$. But then, the
reader would doubtless expect graduates to pair off heterosexually, and
not to arrive randomly for lunch. Commonsense is a great asset.

4. Of the ships "sampled" at each engagement, some were captured, the
rest not. We should reconstruct the table as to give the numbers in each
category and the sample totals. Thus:

	Battle				
	A	B	C	D	Total
Taken	7	4	8	20	39
Escaped	19	23	8	13	63
Total	26	27	16	33	102

(7.2) enables us to fill in the required E_{ij} values, which it were sensible
to round to the nearest unit; since no significance can be attached to

capturing 0.324 of a ship. We compute χ^2 using (7.3). There are $(4 - 1)(2 - 1) = 3$ d.f. Doing the arithmetic yields $\chi^2 = 14.47$; whence $p(H_0) < 0.01$. Even if we were fashionably sceptical, and had set an α of 0.01, we would have to conclude that there really was something rather special about H. Nelson.

Chapter 8

1. Clearly, the hypothesis under test is that there is a difference between learning motivated by Threat of Punishment (P) and that motivated by Promise of Reward (R). The null hypothesis H_0 is that there is no difference; or, at least, that none is likely to show up in the present case. As hardened cynics in these matters, we will take a lot of convincing to reject H_0; and we set α at 0.025.

   This is a good case for U, and we rank the table easily enough:

55	59	61	62	65	75	78	80	85	86	92
R	R	P	P	R	R	P	R	P	P	P

U is obtained by counting the number of Ps preceding each R; thus: $U = 0 + 0 + 2 + 2 + 3 = 7$. Consulting Appendix Table F for $m = 5$, $n = 6$, we find that $p = 0.089$. This is not sufficient warrant to reject H_0.

2. Since each gourmet acted as his own control, the Wilcoxon T-test seems appropriate here. As the matter is (to be honest) of no great moment, we may fix α at 0.10. If we take the differences (X-Score) − (Y-Score), we note that one is zero, reducing N from 10 to 9, and that only two are of negative sign. These have ranks $1\frac{1}{2}$ and 5, giving $T = 6\frac{1}{2}$. Consulting Appendix Table G we find that, for an α of 0.05, this would not quite reject H_0 (two-tailed, of course; we have no prior opinion on the merits of the cooks). However, noticing how the numbers in the table are varying, we presume that H_0 is rejected at our α and award the grand prix de diner to M.X.

3. Here again we have a good case for using the U-test. Having carried out the ranking, we find that $n = m = 36$, that $Rn = 1200$, $Rm = 1428$ whence substitution in (8.2) gives:

$$U = 36 \cdot 36 + \frac{36 \cdot 37}{2} - 1428 = 534$$

We do not need to check whether we have found U in error: since $n = m$, we only have to take the larger R to be sure of having the right value. Since n and m are both > 20, we may transform to the Gaussian variable z, but because of the very large number of tied scores (four of 1, four of 2, seventeen of 3, twenty-one of 4, twelve of 5, ten of 6 and three of 7); we feel obliged to use the slightly more complex form of expression (8.6). Simple arithmetic finds $\Sigma T = 1415.5$ whence substitution in (8.6) gives:

$$z = \frac{534 - \dfrac{36 \cdot 36}{2}}{\sqrt{\left[\dfrac{36^2}{72 \cdot 71} \left(\dfrac{72^3 - 72}{12} - 1415.5 \right) \right]}} \doteq 1.33$$

Appendix Table A gives $p = 0.09$. The reader has not been given enough information to fix an α; but neither my colleague nor I think that H_0 should be rejected on these data.

4. I nowhere said that the twins were *identical*, so the reader has no right to assume that the pairs were matched. Thus Wilcoxon's test seems hardly suitable, though the U-test does. If we carry out the ranking, we note that $n = m = 11$ that the rank total Rn for the "pure" twins is the higher: $Rn = 140$ whence, substituting in (8.2):

$$U = 11 \cdot 11 + \frac{11 \cdot 12}{2} - 140 = 47$$

We have not yet determined α. Now, whatever the FRI might set, we don't believe a word of it, and set $\alpha = 0.01$ (the question is one-tailed, of course). Consulting our tables we find that in fact, the U we have obtained would not suffice to reject H_0 even at the relatively slack α of 0.05. All that educational effort for nothing!

Chapter 9

1. Here we are looking at separate groups of unequal size, to see whether they differ in the measured quality. Clearly, the Kruskal—Wallis test is the one to use. Equally clearly, H_0 is that the groups do not differ. It is not so easy, however, to assign α: we might be very sceptical, and insist on a value of 0.01; while Aristotle and his pupils would be nearly convinced already, and willing to accept 0.10. We might compromise on 0.05.

Ranking all the scores, computing the several T_i, we can substitute in (9.2) to find H:

$$H = \frac{12}{34 \cdot 35} \left(\frac{134^2}{6} + \frac{119^2}{5} + \frac{83^2}{6} + \frac{61^2}{5} + \frac{91^2}{5} + \frac{107^2}{6} \right) - 3.35 \doteq 7.5$$

The numbers being large, we may test this as a value of χ^2 on 5 d.f.; and consulting Appendix Table D we find that $p(H_0)$ is rather nearer to 0.2 than to 0.1. Racists retire, gnashing their teeth.

2. Equation (9.1) states:

$$H = \frac{12}{N(N+1)} \sum_{i=1}^{k} R_i \left(\bar{T}_i - \frac{N+1}{2} \right)^2$$

$$= \frac{12}{N(N+1)} \left[\sum_{i=1}^{k} R_i \bar{T}_i^2 - (N+1) \sum_{i=1}^{k} R_i \bar{T}_i + (N+1)^2 \sum_{i=1}^{k} R_i \right]$$

$$= \frac{12}{N(N+1)} \left[\sum_{i=1}^{k} \frac{T_i^2}{R_i} - (N+1) \sum_{i=1}^{k} T_i + \frac{N(N+1)^2}{4} \right]$$

$$= \frac{12}{N(N+1)} \left[\sum_{i=1}^{k} \frac{T_i^2}{R_i} - \frac{(N+1)N(N+1)}{2} + \frac{N(N+1)^2}{4} \right]$$

$$= \frac{12}{N(N+1)} \left[\sum_{i=1}^{k} \frac{T_i^2}{R_i} - \frac{N(N+1)^2}{4} \right]$$

$$= \frac{12}{N(N+1)} \left[\sum_{i=1}^{k} \frac{T_i^2}{R_i} \right] - 3(N+1)$$

which is equation (9.2).

3. In this problem, the scores may be taken as genuinely matched, and consequently the Friedman test is entirely appropriate. There is no difficulty with H_0: it is that the additives do not produce any noticeable difference in heights. Since caution is appropriate in all matters of this kind, we would do well to fix a fairly strict α of 0.01. Since $n = 10$, we must use (9.5) to find χ^2, which we readily do, finding that

$$\chi^2 = \frac{12}{120} (324 + 361 + 529) - 120$$

$$\doteq 121.4 - 120$$

$$= 1.4$$

This is equivalent to χ^2 on 2 d.f., which shows that $p(H_0) = 0.5$. There seems to be nothing in it.

A point worth noticing here is that the range of row medians is 15 cm, whereas the maximum difference of values *within* any row is 3 cm. Now, although the present result is negative, it were entirely possible to obtain a highly "significant" result under exactly these limits; which demonstrates the important point that a "significant" difference, statistically speaking, may contribute but little of "significance", as generally used, to population variability.

4. Again, this is evidently a case for the Kruskal–Wallis test, with H_0 being that there is no difference in credibility between the several liars. We

may suppose that even a small difference here could make a large monetary difference to the agent's employers; so we may set α at 0.05. There are many tied ranks in the table; so we must use the corrected expression (9.3) for H'.

First using (9.2) to find H, we obtain:

$$H = 0.01$$

There is no point in carrying out the correction, although the arithmetic is not difficult. There is nothing in this. It does *not* follow, however, that the liars are equally usable by the advertisers; they might vary in their capacity to draw crowds, in the fees they charged, or in a number of other ways.

Chapter 10

1. The question at issue here is not "Do the values obtained by the two experts from each corpse differ?" It is evident that in eight cases out of nine they do. The question is "Is the mean difference between the two experts' estimates notably different from zero?" We must therefore produce a new table giving the *differences* and subject this sample of differences to test, as to whether its mean is different from zero. The new table is:

Corpse	1	2	3	4	5	6	7	8	9
Difference of estimates (mg)	0	−1.0	−1.5	0.2	−0.1	1.6	−0.3	0.4	0.9

The mean of the sample $= \bar{x} = 0.02$. This is very small, but the QC is very keen on a result, and would settle for an α of 0.1 ("Gentlemen of the Jury, there is less than one chance in ten . . ."), so we have to do the sums. We find that $S_1^2 = 0.87$, so substituting in (10.2) we obtain:

$$t = \frac{0.2 \times 3}{\sqrt{(0.87)}} \doteqdot 0.07$$

There are 8 d.f.; so the table shows that we cannot reject H_0. The QC must find some other ploy.

2. The mean mass of cakes in batch A $= \bar{x} = 240.7$ g, that of cakes in batch B $= \bar{y} = 245.5$ g. Hence:

$$(\bar{x} - \bar{y}) = 4.8 \text{ g}$$

This is not large, but could make a very substantial difference to the manufacturer's costs over tens of thousands of cakes. The inspector, therefore, accepts a traditional α of 0.05, and carries out the sums. We find that the s.d.s of the two samples are similar enough; so substituting in (10.3) we obtain:

$$t = \frac{4.8}{\sqrt{\left(\frac{4}{3} \cdot \frac{2}{10}\right)}} = \frac{4.8}{\sqrt{(0.267)}} \doteqdot 9.3$$

Which is sufficient to reject H_0 at the chosen α. The inspector must inform the management at once.

3. The mean distance covered by the eleven cars in the sample is 47.52 km, that is to say, the mean difference from 50.00 km is 2.48 km. The salesman, though honest, does not seek a row with the manufacturers, and therefore sets a fairly strict α of 0.01. Substituting in (10.2) gives

$$t = \frac{2.48}{\sqrt{(0.225)}} \doteqdot 5.2$$

This is quite sufficient to reject H_0 at the chosen α. However, it does *not*, by itself, prove that the manufacturers were mendacious. It remains to be proved that the dealer's sample was indeed a random one; which might not be the case. Still, there is definitely something to be looked into.

4. Long before the RSPCA arrives to make a fuss, we have noticed that, not only are the two samples thoroughly lopsided, but they are lopsided in opposite directions. Therefore the use of t is not valid.

Chapter 11

1. Substituting in equations (11.1)–(11.4) and rounding to the nearest unit are found the following:

	Variance estimate	Variance ratio
subjects	525	21
tests	147	6
residual	25	

Appendix Table L shows that the important source of variation is that between subjects (an α of 0.01 would be satisfied); the rest seem trivial.

2. Substituting in (11.5) and rounding sums of sqs to the nearest unit, we find:

Source	Sum of sqs	d.f.	V.E.	V.R.
times	240	3	80	10.7
tests	39	3	13	1.7
subjects	897	3	299	40
R	45	6		
total	1221	15		

Consulting Appendix Table L, we find that time is an important source of variation (an α of 0.01 would be satisfied); but that it is less so than the variability of subjects.

3. What the investigator needs to know is: is there an important source of variability which he must seek to eliminate before introducing noise as his experimental variable? We first notice (or ought to), on looking at the numbers, that the overall variability is not large, considering the size of the scores. Arithmetic (substituting in (11.5)) confirms this. The variance ratios for tests, order and subjects are approximately 8, 5 and 1.2 respectively, none of which would satisfy an α of 0.05. The experimenter may heave a sigh of relief and go ahead.

Chapter 12

1. It is easy to rank the two time-series given by the politician; the sequence for numbers of purple immigrants is the series of natural numbers 1−8, and the corresponding one for convictions for brothel keeping is 1, 3, 2, 4, 5, 7, 6, 8. The simple application of (12.9) gives $\rho = 0.95$. I have no idea what the politician's α might be (very slack, I fancy) but this would suffice to reject H_0 at an α of 0.01.

And so what? We might well find a similar (or better) correlation with, e.g., the number of privately owned cars. In the absence of an external reason to expect such a relationship, the politician is only "blinding with stats".

2. Recalling that degree of confidence is *greatest* when the rating is *smallest* (1 = certain) we may produce a rank table thus

	S								
	1	2	3	4	5	6	7	8	9
C-rank	8	1	5½	3½	8	2	8	3½	5½
E-rank	8	1	5½	4	7	2	9	3	5½

which readily yields a $\Sigma\, d^2$ value of 2.5 (if we are going to use ρ). However, there are rather a lot of tied ranks, so we ought to use the corrected expression (12.11). Substituting therein we obtain:

$$\rho = \frac{57 + 59.5 - 2.5}{2\sqrt{(57 \cdot 59.5)}} = \frac{57}{\sqrt{(3391.5)}}$$

$$\doteq 0.98$$

which would reject H_0 at an α of 0.01. I must admit, however, that I am not sure about what to make of this. It is the general experience of mankind that confidence is generally correlated with error; and this seems to be a confirmatory datum: to say more than that would be unwise.

3. This is a completely straightforward question: is the concordance between the several rank orders for subjects better than random? Since, on general grounds, we might well expect it to be, we could reasonably set an α of 0.05 and proceed, perhaps a shade laboriously but without real difficulty, to rank the six sets of marks and compute W. This we find to be 0.913; but since $M > 7$, we convert this to χ^2 (using (12.18)) obtaining a value of 76.7 with 14 d.f. This amply satisfies our criterion, for $p(H_0) < 0.001$. Nobody is surprised, and we go on our way rejoicing.

4. The data being ready ranked, it is a small thing to find that $W = 0.01$ and that concordance is negligible. It is also a damn stupid thing to do. If, instead of rushing to our formulae like well-drilled parrots, we had *looked at the numbers*, as good statisticians always should, we would have noticed that the judges seem to fall into two subsets of three. These subjects appear to contradict one another, but to agree internally, as we see from Table A.1.

Table A.1 Ranking of eight paintings by two sets of three Judges (data from Table 12.E4)

		Painting							
		a	b	c	d	e	f	g	h
(*a*)	Judge 1	3	1	4	5	2	6	8	7
	3	3	1	4	5	2	6	7	8
	4	4	3	2	4	6	1	5	8	7
	R_i	9	4	12	16	15	17	23	22
(*b*)	Judge 2	6	8	4	5	7	3	2	1
	5	5	6	7	4	5	8	3	1	2
	6	6	5	8	6	4	7	3	2	1
	R_i	17	23	14	14	22	9	5	4

If we now examine these subsets severally, we find that Judges 1, 3 and 4 yield a W of 0.968, hence a χ^2 (7 d.f.) of 20.3; while the corresponding values for Judges 2, 5 and 6 are $W = 0.947$ and $\chi^2 = 19.9$. In both cases

$p(H_0) < 0.01$ and we could rank the paintings according to the two subsets thus:

Subset (a)	b	e	a	c	d	f	h	g
Subset (b)	h	g	f	c & d		a	e	b

which is as near to flat disagreement as you are likely to find. Possibly the two subsets were "ancients" and "moderns".

The moral is: never fail to take a good long look at the numbers.

Appendix: Tables A–O

Table A   The Gaussian distribution

The table gives proportions of area under the curve of the unit Gaussian distribution whose mean is 0 and standard deviation 1.

z	Area between mean and z	z	Area between mean and z	z	Area between mean and z
0.00	0.0000	0.20	0.0793	0.40	0.1554
0.01	0.0040	0.21	0.0832	0.41	0.1591
0.02	0.0080	0.22	0.0871	0.42	0.1628
0.03	0.0120	0.23	0.0910	0.43	0.1664
0.04	0.0160	0.24	0.0948	0.44	0.1700
0.05	0.0199	0.25	0.0987	0.45	0.1736
0.06	0.0239	0.26	0.1026	0.46	0.1772
0.07	0.0279	0.27	0.1064	0.47	0.1808
0.08	0.0319	0.28	0.1103	0.48	0.1844
0.09	0.0359	0.29	0.1141	0.49	0.1879
0.10	0.0398	0.30	0.1179	0.50	0.1915
0.11	0.0438	0.31	0.1217	0.51	0.1950
0.12	0.0478	0.32	0.1255	0.52	0.1985
0.13	0.0517	0.33	0.1293	0.53	0.2019
0.14	0.0557	0.34	0.1331	0.54	0.2054
0.15	0.0596	0.35	0.1368	0.55	0.2088
0.16	0.0636	0.36	0.1406	0.56	0.2123
0.17	0.0675	0.37	0.1443	0.57	0.2157
0.18	0.0714	0.38	0.1480	0.58	0.2190
0.19	0.0753	0.39	0.1517	0.59	0.2224

Table A — *continued*

z	Area between mean and z	z	Area between mean and z	z	Area between mean and z
0.60	0.2257	0.90	0.3159	1.20	0.3849
0.61	0.2291	0.91	0.3186	1.21	0.3869
0.62	0.2324	0.92	0.3212	1.22	0.3888
0.63	0.2357	0.93	0.3238	1.23	0.3907
0.64	0.2389	0.94	0.3264	1.24	0.3925
0.65	0.2422	0.95	0.3289	1.25	0.3944
0.66	0.2454	0.96	0.3315	1.26	0.3962
0.67	0.2486	0.97	0.3340	1.27	0.3980
0.68	0.2517	0.98	0.3365	1.28	0.3997
0.69	0.2549	0.99	0.3389	1.29	0.4015
0.70	0.2580	1.00	0.3413	1.30	0.4032
0.71	0.2611	1.01	0.3438	1.31	0.4049
0.72	0.2642	1.02	0.3461	1.32	0.4066
0.73	0.2673	1.03	0.3485	1.33	0.4082
0.74	0.2704	1.04	0.3508	1.34	0.4099
0.75	0.2734	1.05	0.3531	1.35	0.4115
0.76	0.2764	1.06	0.3554	1.36	0.4131
0.77	0.2794	1.07	0.3577	1.37	0.4147
0.78	0.2823	1.08	0.3599	1.38	0.4162
0.79	0.2852	1.09	0.3621	1.39	0.4177
0.80	0.2881	1.10	0.3643	1.40	0.4192
0.81	0.2910	1.11	0.3665	1.41	0.4207
0.82	0.2939	1.12	0.3686	1.42	0.4222
0.83	0.2967	1.13	0.3708	1.43	0.4236
0.84	0.2995	1.14	0.3729	1.44	0.4251
0.85	0.3023	1.15	0.3749	1.45	0.4265
0.86	0.3051	1.16	0.3770	1.46	0.4279
0.87	0.3078	1.17	0.3790	1.47	0.4292
0.88	0.3106	1.18	0.3810	1.48	0.4306
0.89	0.3133	1.19	0.3830	1.49	0.4319

Table A — *continued*

z	Area between mean and z	z	Area between mean and z	z	Area between mean and z
1.50	0.4332	1.80	0.4641	2.10	0.4821
1.51	0.4345	1.81	0.4649	2.11	0.4826
1.52	0.4357	1.82	0.4656	2.12	0.4830
1.53	0.4370	1.83	0.4664	2.13	0.4834
1.54	0.4382	1.84	0.4671	2.14	0.4838
1.55	0.4394	1.85	0.4678	2.15	0.4842
1.56	0.4406	1.86	0.4686	2.16	0.4846
1.57	0.4418	1.87	0.4693	2.17	0.4850
1.58	0.4429	1.88	0.4699	2.18	0.4854
1.59	0.4441	1.89	0.4706	2.19	0.4857
1.60	0.4452	1.90	0.4713	2.20	0.4861
1.61	0.4463	1.91	0.4719	2.21	0.4864
1.62	0.4474	1.92	0.4726	2.22	0.4868
1.63	0.4484	1.93	0.4732	2.23	0.4871
1.64	0.4495	1.94	0.4738	2.24	0.4875
1.65	0.4505	1.95	0.4744	2.25	0.4878
1.66	0.4515	1.96	0.4750	2.26	0.4881
1.67	0.4525	1.97	0.4756	2.27	0.4884
1.68	0.4535	1.98	0.4761	2.28	0.4887
1.69	0.4545	1.99	0.4767	2.29	0.4890
1.70	0.4554	2.00	0.4772	2.30	0.4893
1.71	0.4564	2.01	0.4778	2.31	0.4896
1.72	0.4573	2.02	0.4783	2.32	0.4898
1.73	0.4582	2.03	0.4788	2.33	0.4901
1.74	0.4591	2.04	0.4793	2.34	0.4904
1.75	0.4599	2.05	0.4798	2.35	0.4906
1.76	0.4608	2.06	0.4803	2.36	0.4909
1.77	0.4616	2.07	0.4808	2.37	0.4911
1.78	0.4625	2.08	0.4812	2.38	0.4913
1.79	0.4633	2.09	0.4817	2.39	0.4916

Table A — *continued*

z	Area between mean and z	z	Area between mean and z	z	Area between mean and z
2.40	0.4918	2.72	0.4967	3.04	0.4988
2.41	0.4920	2.73	0.4968	3.05	0.4989
2.42	0.4922	2.74	0.4969	3.06	0.4989
2.43	0.4925	2.75	0.4970	3.07	0.4989
2.44	0.4927	2.76	0.4971	3.08	0.4990
2.45	0.4929	2.77	0.4972	3.09	0.4990
2.46	0.4931	2.78	0.4973	3.10	0.4990
2.47	0.4932	2.79	0.4974	3.11	0.4991
2.48	0.4934	2.80	0.4974	3.12	0.4991
2.49	0.4936	2.81	0.4975	3.13	0.4991
2.50	0.4938	2.82	0.4976	3.14	0.4992
2.51	0.4940	2.83	0.4977	3.15	0.4992
2.52	0.4941	2.84	0.4977	3.16	0.4992
2.53	0.4943	2.85	0.4978	3.17	0.4992
2.54	0.4945	2.86	0.4979	3.18	0.4993
2.55	0.4946	2.87	0.4979	3.19	0.4993
2.56	0.4948	2.88	0.4980	3.20	0.4993
2.57	0.4949	2.89	0.4981	3.21	0.4993
2.58	0.4951	2.90	0.4981	3.22	0.4994
2.59	0.4952	2.91	0.4982	3.23	0.4994
2.60	0.4953	2.92	0.4982	3.24	0.4994
2.61	0.4955	2.93	0.4983	3.25	0.4994
2.62	0.4956	2.94	0.4984	3.30	0.4995
2.63	0.4957	2.95	0.4984	3.35	0.4996
2.64	0.4959	2.96	0.4985	3.40	0.4997
2.65	0.4960	2.97	0.4985	3.45	0.4997
2.66	0.4961	2.98	0.4986	3.50	0.4998
2.67	0.4962	2.99	0.4986	3.60	0.4998
2.68	0.4963	3.00	0.4987	3.70	0.4999
2.69	0.4964	3.01	0.4987	3.80	0.4999
2.70	0.4965	3.02	0.4987	3.90	0.4995
2.71	0.4966	3.03	0.4988	4.00	0.4997

Table B   Random numbers

93 47 43 73 86	36 96 47 36 61	46 98 63 71 62	33 26 16 80 45	60 11 14 10 95
97 74 24 67 62	42 81 14 57 20	42 53 32 37 32	27 07 36 07 51	24 51 79 89 73
16 76 62 27 66	56 50 26 71 07	32 90 79 78 53	13 55 38 58 59	88 97 54 14 10
12 56 85 99 26	96 96 68 27 31	05 03 72 93 15	57 12 10 14 21	88 26 49 81 76
55 59 56 35 64	38 54 82 46 22	31 62 43 09 90	06 18 44 32 53	23 83 01 30 30
16 22 77 94 39	49 54 43 54 82	17 37 93 23 78	87 35 20 96 43	84 26 34 91 64
84 42 17 53 31	57 24 55 06 88	77 04 74 47 67	21 76 33 50 25	83 92 12 06 76
63 01 63 78 59	16 95 55 67 19	98 10 50 71 75	12 86 73 58 07	44 39 52 38 79
33 21 12 34 29	78 64 56 07 82	52 42 07 44 38	15 51 00 13 42	99 66 02 79 54
57 60 86 32 44	09 47 27 96 54	49 17 46 09 62	90 52 84 77 27	08 02 73 43 28
18 18 07 92 46	44 17 16 58 09	79 83 86 19 62	06 76 50 03 10	55 23 64 05 05
26 62 38 97 75	84 16 07 44 99	83 11 46 32 24	20 14 85 88 45	10 93 72 88 71
23 42 40 64 74	82 97 77 77 81	07 45 32 14 08	32 98 94 07 72	93 85 79 10 75
82 36 28 19 95	50 92 26 11 97	00 56 76 31 38	80 22 02 53 53	86 60 42 04 53
37 85 94 35 12	83 39 50 08 30	42 34 07 96 88	54 42 06 87 98	35 85 29 48 39
70 29 17 12 13	40 33 20 38 26	13 89 51 03 74	17 76 37 13 04	07 74 21 19 30
56 62 18 37 35	96 83 50 87 75	97 12 25 93 47	70 33 24 03 54	97 77 46 44 80
99 49 57 22 77	88 42 95 45 72	16 64 36 16 00	04 43 18 66 79	94 77 24 21 90
16 08 15 04 72	33 27 14 34 09	45 59 34 68 49	12 72 07 34 45	99 27 72 95 14
31 16 93 32 43	50 27 89 87 19	20 15 37 00 49	52 85 66 60 44	38 68 88 11 80
68 34 30 13 70	55 74 30 77 40	44 22 78 84 26	04 33 46 09 52	68 07 97 06 57
74 57 25 65 76	59 29 97 68 60	71 91 38 67 54	13 58 18 24 76	15 54 55 95 52
27 42 37 86 53	48 55 90 65 72	96 57 69 36 10	96 46 92 42 45	97 60 49 04 91
00 39 68 29 61	66 37 32 20 30	77 84 57 03 29	10 45 65 04 26	11 04 96 67 24
29 94 98 94 24	68 49 69 10 82	53 75 91 93 30	34 25 20 57 27	40 48 73 51 92
16 90 82 66 59	83 62 64 11 12	67 19 00 71 74	60 47 21 29 68	02 02 37 03 31
11 27 94 75 06	06 09 19 74 66	02 94 37 34 02	76 70 90 30 86	38 45 94 30 38
35 24 10 16 20	33 32 51 26 38	79 78 45 04 91	16 92 53 56 16	02 75 50 95 98
38 23 16 86 38	42 38 97 01 50	87 75 66 81 41	40 01 74 91 62	48 51 84 08 32
31 96 25 91 47	96 44 33 49 13	34 86 82 53 91	00 52 43 48 85	27 55 26 89 62
56 67 40 67 14	64 05 71 95 86	11 05 65 09 68	76 83 20 37 90	57 16 00 11 66
14 90 84 45 11	75 73 88 05 90	52 27 41 14 86	22 98 12 22 08	07 52 74 95 80
68 05 51 18 00	33 96 02 75 19	07 60 62 93 55	59 33 82 43 90	49 37 38 44 59
20 46 78 73 90	97 51 40 14 02	04 02 33 31 08	39 54 16 49 36	47 95 93 13 30
64 19 58 97 79	15 06 15 93 20	01 90 10 75 06	40 78 78 89 62	02 67 74 17 33
05 26 93 70 60	22 35 85 15 13	92 03 51 59 77	59 56 78 06 83	52 91 05 70 74
07 97 10 88 23	09 98 42 99 64	61 71 62 99 15	06 51 29 16 93	58 05 77 09 51
68 71 86 85 85	54 87 66 47 54	73 32 08 11 12	44 95 92 63 16	29 56 24 29 48
26 99 61 65 53	58 37 78 80 70	42 10 50 67 42	32 17 55 85 74	94 44 67 16 94
14 65 52 68 75	87 59 36 22 41	26 78 63 06 55	13 08 27 01 50	15 29 39 39 43
17 53 77 58 71	71 41 61 50 72	12 41 94 96 26	44 95 27 36 99	02 96 74 30 83
90 26 59 21 19	23 52 23 33 12	96 93 02 18 39	07 02 18 36 07	25 99 32 70 23
41 23 52 55 99	31 04 49 69 96	10 47 48 45 88	13 41 43 89 20	97 17 14 49 17
60 20 50 81 69	31 99 73 68 68	35 81 33 03 76	24 30 12 48 60	18 99 10 72 34
91 25 38 05 90	94 58 28 41 36	45 37 59 03 09	90 35 57 29 12	82 62 54 65 60
34 50 57 74 37	98 80 33 00 91	09 77 93 19 82	74 94 80 04 04	45 07 31 66 49
35 22 04 39 43	73 81 53 94 79	33 62 46 86 28	08 31 54 46 31	53 94 13 38 47
09 79 13 77 48	73 82 97 22 21	05 03 27 24 83	72 89 44 05 60	35 80 39 94 88
88 75 80 18 14	22 95 75 42 49	39 32 82 22 49	02 48 07 70 37	16 04 61 67 87
90 96 23 70 00	39 00 03 06 90	55 85 78 38 36	94 37 30 69 32	90 89 00 76 33

Table C   The binomial test

The table gives the probabilities of observed values as small as x (the smaller of the frequencies) supposing that there is a one-tailed question and that H_0 is $p = q = 0.5$. Decimal points have been left out.

N	0	1	2	3	4	5	6	7	8	9	10	11	12	13	14	15
5	031	188	500	812	969	†										
6	016	109	344	656	891	984	†									
7	008	062	227	500	773	938	992	†								
8	004	035	145	363	637	855	965	996	†							
9	002	020	090	254	500	746	910	980	998	†						
10	001	011	055	172	377	623	828	945	989	999	†					
11		006	033	113	274	500	726	887	967	994	†		†			
12		003	019	073	194	387	613	806	927	981	997	†		†		
13		002	011	046	133	291	500	709	867	954	989	998	†		†	
14		001	006	029	090	212	395	605	788	910	971	994	999	†		†
15			004	018	059	151	304	500	696	849	941	982	996	†	†	†
16			002	011	038	105	227	402	598	773	895	962	989	998	†	†
17			001	006	025	072	166	315	500	685	834	928	975	994	999	†
18			001	004	015	048	119	240	407	593	760	881	952	985	996	999
19				002	010	032	084	180	324	500	676	820	916	968	990	998
20				001	006	021	058	132	252	412	588	748	868	942	979	994
21				001	004	013	039	095	192	332	500	668	808	905	961	987
22					002	008	026	067	143	262	416	584	738	857	933	974
23					001	005	017	047	105	202	339	500	661	798	895	953
24					001	003	011	032	076	154	271	419	581	729	846	924
25						002	007	022	054	115	212	345	500	655	788	885

Table D The χ^2 test

The table gives the probability, given H_0 and n d.f. of obtained values of χ^2 as large as, or larger than those given (e.g., for 5 d.f. there is a probability of 0.02 of obtaining a $\chi^2 \geqslant 13.388$).

n	0.99	0.98	0.95	0.90	0.80	0.70	0.50	0.30	0.20	0.10	0.05	0.02	0.01	0.001
1	0.0^3157	0.0^6628	0.00393	0.0158	0.0642	0.148	0.455	1.074	1.642	2.706	3.841	5.412	6.635	10.827
2	0.0201	0.0404	0.103	0.211	0.446	0.713	1.386	2.408	3.219	4.605	5.991	7.824	9.210	13.815
3	0.115	0.185	0.352	0.584	1.005	1.424	2.366	3.665	4.642	6.251	7.815	9.837	11.345	16.266
4	0.297	0.429	0.711	1.064	1.649	2.195	3.357	4.878	5.989	7.779	9.488	11.668	13.277	18.467
5	0.554	0.752	1.145	1.610	2.343	3.000	4.351	6.064	7.289	9.236	11.070	13.388	15.086	20.515
6	0.872	1.134	1.635	2.204	3.070	3.828	5.348	7.231	8.558	10.645	12.592	15.033	16.812	22.457
7	1.239	1.564	2.167	2.833	3.822	4.671	6.346	8.383	9.803	12.017	14.067	16.622	18.475	24.322
8	1.646	2.032	2.733	3.490	4.594	5.527	7.344	9.524	11.030	13.362	15.507	18.168	20.090	26.125
9	2.088	2.532	3.325	4.168	5.380	6.393	8.343	10.656	12.242	14.684	16.919	19.679	21.666	27.877
10	2.558	3.059	3.940	4.865	6.179	7.267	9.342	11.781	13.442	15.987	18.307	21.161	23.209	29.588
11	3.053	3.609	4.575	5.578	6.989	8.148	10.341	12.899	14.631	17.275	19.675	22.618	24.725	31.264
12	3.571	4.178	5.226	6.304	7.807	9.034	11.340	14.011	15.812	18.549	21.026	24.054	26.217	32.909
13	4.107	4.765	5.892	7.042	8.634	9.926	12.340	15.119	16.985	19.812	22.362	25.472	27.688	34.528
14	4.660	5.368	6.571	7.790	9.467	10.821	13.339	16.222	18.151	21.064	23.685	26.873	29.141	36.123
15	5.229	5.985	7.261	8.547	10.307	11.721	14.339	17.322	19.311	22.307	24.996	28.259	30.578	37.697
16	5.812	6.614	7.962	9.312	11.152	12.624	15.338	18.418	20.465	23.542	26.296	29.633	32.000	39.252
17	6.408	7.255	8.672	10.085	12.002	13.531	16.338	19.511	21.615	24.769	27.587	30.995	33.409	40.790
18	7.015	7.906	9.390	10.865	12.857	14.440	17.338	20.601	22.760	25.989	28.869	32.346	34.805	42.312
19	7.633	8.567	10.117	11.651	13.716	15.352	18.338	21.689	23.900	27.204	30.144	33.687	36.191	43.820
20	8.260	9.237	10.851	12.443	14.578	16.266	19.337	22.775	25.038	28.412	31.410	35.020	37.566	45.315
21	8.897	9.915	11.591	13.240	15.445	17.182	20.337	23.858	26.171	29.615	32.671	36.343	38.932	46.797
22	9.542	10.600	12.338	14.041	16.314	18.101	21.337	24.939	27.301	30.813	33.924	37.659	40.289	48.268
23	10.196	11.293	13.091	14.848	17.187	19.021	22.337	26.018	28.429	32.007	35.172	38.968	41.638	49.728
24	10.856	11.992	13.848	15.659	18.062	19.943	23.337	27.096	29.553	33.196	36.415	40.270	42.980	51.179
25	11.524	12.697	14.611	16.473	18.940	20.867	24.337	28.172	30.675	34.382	37.652	41.566	44.314	52.620
26	12.198	13.409	15.379	17.292	19.820	21.792	25.336	29.246	31.795	35.563	38.885	42.856	45.642	54.052
27	12.879	14.125	16.151	18.114	20.703	22.719	26.336	30.319	32.912	36.741	40.113	44.140	46.963	55.476
28	13.565	14.847	16.928	18.939	21.588	23.647	27.336	31.391	34.027	37.916	41.337	45.419	48.278	56.893
29	14.256	15.574	17.708	19.768	22.475	24.577	28.336	32.461	35.139	39.087	42.557	46.693	49.488	58.302
30	14.953	16.306	18.493	20.599	23.364	25.508	29.336	33.530	36.250	40.256	43.773	47.962	50.892	59.703

Tables E$_1$ and E$_2$ The "runs" test

The tables give critical values of r for a range of values of n_1 and n_2. Any value of r equal to or less than that given in Table E$_1$, or equal to or greater than that given in Table E$_2$, enables H_0 to be rejected at an α of 0.05.

E$_1$

n_1 \\ n_2	2	3	4	5	6	7	8	9	10	11	12	13	14	15	16	17	18	19	20
2											2	2	2	2	2	2	2	2	2
3			2	2	2	2	2	2	2	2	2	3	3	3	3	3	3	3	3
4			2	2	2	3	3	3	3	3	3	3	3	4	4	4	4	4	4
5			2	2	3	3	3	3	3	4	4	4	4	4	4	4	5	5	5
6		2	2	3	3	3	3	4	4	4	4	5	5	5	5	5	5	6	6
7		2	2	3	3	3	4	4	5	5	5	5	5	6	6	6	6	6	6
8		2	3	3	3	4	4	5	5	5	6	6	6	6	6	7	7	7	7
9		2	3	3	4	4	5	5	5	6	6	6	7	7	7	7	8	8	8
10		2	3	3	4	5	5	5	6	6	7	7	7	7	8	8	8	8	9
11		2	3	4	4	5	5	6	6	7	7	7	8	8	8	9	9	9	9
12	2	2	3	4	4	5	6	6	7	7	7	8	8	8	9	9	9	10	10
13	2	2	3	4	5	5	6	6	7	7	8	8	9	9	9	10	10	10	10
14	2	2	3	4	5	5	6	7	7	8	8	9	9	9	10	10	10	11	11
15	2	3	3	4	5	6	6	7	7	8	8	9	9	10	10	11	11	11	12
16	2	3	4	4	5	6	6	7	8	8	9	9	10	10	11	11	11	12	12
17	2	3	4	4	5	6	7	7	8	9	9	10	10	11	11	11	12	12	13
18	2	3	4	5	5	6	7	8	8	9	9	10	10	11	11	12	12	13	13
19	2	3	4	5	6	6	7	8	8	9	10	10	11	11	12	12	13	13	13
20	2	3	4	5	6	6	7	8	9	9	10	10	11	12	12	13	13	13	14

E$_2$

n_1 \\ n_2	2	3	4	5	6	7	8	9	10	11	12	13	14	15	16	17	18	19	20
2																			
3																			
4				9	9														
5			9	10	10	11	11												
6			9	10	11	12	12	13	13	13	13								
7				11	12	13	13	14	14	14	14	15	15	15					
8				11	12	13	14	14	15	15	16	16	16	16	17	17	17	17	17
9					13	14	14	15	16	16	16	17	17	18	18	18	18	18	18
10					13	14	15	16	16	17	17	18	18	18	19	19	19	20	20
11					13	14	15	16	17	17	18	19	19	19	20	20	20	21	21
12					13	14	16	16	17	18	19	19	20	20	21	21	21	22	22
13						15	16	17	18	19	19	20	20	21	21	22	22	23	23
14						15	16	17	18	19	20	20	21	22	22	23	23	23	24
15						15	16	18	18	19	20	21	22	22	23	23	24	24	25
16							17	18	19	20	21	21	22	23	23	24	25	25	25
17							17	18	19	20	21	22	23	23	24	25	25	26	26
18							17	18	19	20	21	22	23	24	25	25	26	26	27
19							17	18	20	21	22	23	23	24	25	26	26	27	27
20							17	18	20	21	22	23	24	25	25	26	27	27	28

Table F_1 and F_2 The U-test

The first table gives critical values of U for a one-tailed question and $\alpha = 0.025$, or for a two-tailed question and $\alpha = 0.05$.

The second table gives corresponding values for a one-tailed question and $\alpha = 0.01$, or for a two-tailed question and $\alpha = 0.02$.

F_1

n_1	n_2											
	9	10	11	12	13	14	15	16	17	18	19	20
1												
2	0	0	0	1	1	1	1	1	2	2	2	2
3	2	3	3	4	4	5	5	6	6	7	7	8
4	4	5	6	7	8	9	10	11	11	12	13	13
5	7	8	9	11	12	13	14	15	17	18	19	20
6	10	11	13	14	16	17	19	21	22	24	25	27
7	12	14	16	18	20	22	24	26	28	30	32	34
8	15	17	19	22	24	26	29	31	34	36	38	41
9	17	20	23	26	28	31	34	37	39	42	45	48
10	20	23	26	29	33	36	39	42	45	48	52	55
11	23	26	30	33	37	40	44	47	51	55	58	62
12	26	29	33	37	41	45	49	53	57	61	65	69
13	28	33	37	41	45	50	54	59	63	67	72	76
14	31	36	40	45	50	55	59	64	67	74	78	83
15	34	39	44	49	54	59	64	70	75	80	85	90
16	37	42	47	53	59	64	70	75	81	86	92	98
17	39	45	51	57	63	67	75	81	87	93	99	105
18	42	48	55	61	67	74	80	86	93	99	106	112
19	45	52	58	65	72	78	85	92	99	106	113	119
20	48	55	62	69	76	83	90	98	105	112	119	127

Table F − *continued*

F$_2$

n_1	n_2 9	10	11	12	13	14	15	16	17	18	19	20
1												
2					0	0	0	0	0	0	1	1
3	1	1	1	2	2	2	3	3	4	4	4	5
4	3	3	4	5	5	6	7	7	8	9	9	10
5	5	6	7	8	9	10	11	12	13	14	15	16
6	7	8	9	11	12	13	15	16	18	19	20	22
7	9	11	12	14	16	17	19	21	23	24	26	28
8	11	13	15	17	20	22	24	26	28	30	32	34
9	14	16	18	21	23	26	28	31	33	36	38	40
10	16	19	22	24	27	30	33	36	38	41	44	47
11	18	22	25	28	31	34	37	41	44	47	50	53
12	21	24	28	31	35	38	42	46	49	53	56	60
13	23	27	31	35	39	43	47	51	55	59	63	67
14	26	30	34	38	43	47	51	56	60	65	69	73
15	28	33	37	42	47	51	56	61	66	70	75	80
16	31	36	41	46	51	56	61	66	71	76	82	87
17	33	38	44	49	55	60	66	71	77	82	88	93
18	36	41	47	53	59	65	70	76	82	88	94	100
19	38	44	50	56	63	69	75	82	88	94	101	107
20	40	47	53	60	67	73	80	87	93	100	107	114

Table G   The Wilcoxon test

Critical values of T in the Wilcoxon Test.

	Level of significance for one-tailed question		
	0.025	0.01	0.005
	Level of significance for two-tailed question		
N	0.05	0.02	0.01
6	1	—	—
7	2	0	—
8	4	2	0
9	6	3	2
10	8	5	3
11	11	7	5
12	14	10	7
13	17	13	10
14	21	16	13
15	25	20	16
16	30	24	19
17	35	28	23
18	40	33	28
19	46	38	32
20	52	43	37
21	59	49	43
22	66	56	49
23	73	62	55
24	81	69	61
25	90	77	68

Table H   The Kruskal–Wallis test (small samples)

The table gives the probabilities associated with values *as large as* obtained values of the Kruskal–Wallis H.

Sample sizes					Sample sizes				
n_1	n_2	n_3	H	p	n_1	n_2	n_3	H	p
2	1	1	2.7000	0.500	4	2	1	4.8214	0.057
								4.5000	0.076
2	2	1	3.6000	0.200				4.0179	0.114
2	2	2	4.5714	0.067	4	2	2	6.0000	0.014
			3.7143	0.200				5.3333	0.033
								5.1250	0.052
3	1	1	3.2000	0.300				4.4583	0.100
								4.1667	0.105
3	2	1	4.2857	0.100					
			3.8571	0.133	4	3	1	5.8333	0.021
								5.2083	0.050
3	2	2	5.3572	0.029				5.0000	0.057
			4.7143	0.048				4.0556	0.093
			4.5000	0.067				3.8889	0.129
			4.4643	0.105					
					4	3	2	6.4444	0.008
3	3	1	5.1429	0.043				6.3000	0.011
			4.5714	0.100				5.4444	0.046
			4.0000	0.129				5.4000	0.051
								4.5111	0.098
3	3	2	6.2500	0.011				4.4444	0.102
			5.3611	0.032					
			5.1389	0.061	4	3	3	6.7455	0.010
			4.5556	0.100				6.7091	0.013
			4.2500	0.121				5.7909	0.046
								5.7273	0.050
3	3	3	7.2000	0.004				4.7091	0.092
			6.4889	0.011				4.7000	0.101
			5.6889	0.029					
			5.6000	0.050	4	4	1	6.6667	0.010
			5.0667	0.086				6.1667	0.022
			4.6222	0.100				4.9667	0.048
								4.8667	0.054
4	1	1	3.5714	0.200				4.1667	0.082
								4.0667	0.102

Table II *continued*

Sample sizes						Sample sizes				
n_1	n_2	n_3	H	p		n_1	n_2	n_3	H	p
4	4	2	7.0364	0.006					4.8711	0.052
			6.8727	0.011					4.0178	0.095
			5.4545	0.046					3.8400	0.123
			5.2364	0.052						
			4.5545	0.098		5	3	2	6.9091	0.009
			4.4455	0.103					6.8218	0.010
									5.2509	0.049
4	4	3	7.1439	0.010					5.1055	0.052
			7.1364	0.011					4.6509	0.091
			5.5985	0.049					4.4945	0.101
			5.5758	0.051						
			4.5455	0.099		5	3	3	7.0788	0.009
			4.4773	0.102					6.9818	0.011
									5.6485	0.049
4	4	4	7.6538	0.008					5.5152	0.051
			7.5385	0.011					4.5333	0.097
			5.6923	0.049					4.4121	0.109
			5.6538	0.054						
			4.6539	0.097		5	4	1	6.9545	0.008
			4.5001	0.104					6.8400	0.011
									4.9855	0.044
5	1	1	3.8571	0.143					4.8600	0.056
									3.9873	0.098
5	2	1	5.2500	0.036					3.9600	0.102
			5.0000	0.048						
			4.4500	0.071		5	4	2	7.2045	0.009
			4.2000	0.095					7.1182	0.010
			4.0500	0.119					5.2727	0.049
									5.2682	0.050
5	2	2	6.5333	0.008					4.5409	0.098
			6.1333	0.013					4.5182	0.101
			5.1600	0.034						
			5.0400	0.056		5	4	3	7.4449	0.010
			4.3733	0.090					7.3949	0.011
			4.2933	0.122					5.6564	0.049
									5.6308	0.050
5	3	1	6.4000	0.012					4.5487	0.099
			4.9600	0.048					4.5231	0.103

Table H — *continued*

Sample sizes					Sample sizes				
n_1	n_2	n_3	H	p	n_1	n_2	n_3	H	p
5	4	4	7.7604	0.009	5	5	3	7.5780	0.010
			7.7440	0.011				7.5429	0.010
			5.6571	0.049				5.7055	0.046
			5.6176	0.050				5.6264	0.051
			4.6187	0.100				4.5451	0.100
			4.5527	0.102				4.5363	0.102
5	5	1	7.3091	0.009	5	5	4	7.8229	0.010
			6.8364	0.011				7.7914	0.010
			5.1273	0.046				5.6657	0.049
			4.9091	0.053				5.6429	0.050
			4.1091	0.086				4.5229	0.099
			4.0364	0.105				4.5200	0.101
5	5	2	7.3385	0.010	5	5	5	8.0000	0.009
			7.2692	0.010				7.9800	0.010
			5.3385	0.047				5.7800	0.049
			5.2462	0.051				5.6600	0.051
			4.6231	0.097				4.5600	0.100
			4.5077	0.100				4.5000	0.102

Table I The Friedman test (small samples)

The table gives probabilities associated with values as large as obtained values of χ^2 in Friedman test

(1) $K = 3$

$N = 2$		$N = 3$		$N = 4$		$N = 5$	
χ_r^2	p	χ_r^2	p	χ_r^2	p	χ_r^2	p
0	1.000	0.000	1.000	0.0	1.000	0.0	1.000
1	0.833	0.667	0.944	0.5	0.931	0.4	0.954
3	0.500	2.000	0.528	1.5	0.653	1.2	0.691
4	0.167	2.667	0.361	2.0	0.431	1.6	0.522
		4.667	0.194	3.5	0.273	2.8	0.367
		6.000	0.028	4.5	0.125	3.6	0.182
				6.0	0.069	4.8	0.124
				6.5	0.042	5.2	0.093
				8.0	0.0046	6.4	0.039
						7.6	0.024
						8.4	0.0085
						10.0	0.00077

$N = 6$		$N = 7$		$N = 8$		$N = 9$	
χ_r^2	p	χ_r^2	p	χ_r^2	p	χ_r^2	p
0.00	1.000	0.000	1.000	0.00	1.000	0.000	1.000
0.33	0.956	0.286	0.964	0.25	0.967	0.222	0.971
1.00	0.740	0.857	0.768	0.75	0.794	0.667	0.814
1.33	0.570	1.143	0.620	1.00	0.654	0.889	0.865
2.33	0.430	2.000	0.486	1.75	0.531	1.556	0.569
3.00	0.252	2.571	0.305	2.25	0.355	2.000	0.398
4.00	0.184	3.429	0.237	3.00	0.285	2.667	0.328
4.33	0.142	3.714	0.192	3.25	0.236	2.889	0.278
5.33	0.072	4.571	0.112	4.00	0.149	3.556	0.187
6.33	0.052	5.429	0.085	4.75	0.120	4.222	0.154
7.00	0.029	6.000	0.052	5.25	0.079	4.667	0.107
8.33	0.012	7.143	0.027	6.25	0.047	5.556	0.069
9.00	0.0081	7.714	0.021	6.75	0.038	6.000	0.057
9.33	0.0055	8.000	0.016	7.00	0.030	6.222	0.048
10.33	0.0017	8.857	0.0084	7.75	0.018	6.889	0.031
12.00	0.00013	10.286	0.0036	9.00	0.0099	8.000	0.019
		10.571	0.0027	9.25	0.0080	8.222	0.016
		11.143	0.0012	9.75	0.0048	8.667	0.010
		12.286	0.00032	10.75	0.0024	9.556	0.0060
		14.000	0.000021	12.00	0.0011	10.667	0.0035
				12.25	0.00086	10.889	0.0029
				13.00	0.00026	11.556	0.0013
				14.25	0.000061	12.667	0.00066
				16.00	0.0000036	13.556	0.00035
						14.000	0.00020
						14.222	0.000097
						14.889	0.000054
						16.222	0.000011
						18.000	0.0000006

Table I – *continued*

(2) $K = 4$

$N = 2$		$N = 3$		$N = 4$			
χ_r^2	p	χ_r^2	p	χ_r^2	p	χ_r^2	p
0.0	1.000	0.2	1.000	0.0	1.000	5.7	0.141
0.6	0.958	0.6	0.958	0.3	0.992	6.0	0.105
1.2	0.834	1.0	0.910	0.6	0.928	6.3	0.094
1.8	0.792	1.8	0.727	0.9	0.900	6.6	0.077
2.4	0.625	2.2	0.608	1.2	0.800	6.9	0.068
3.0	0.542	2.0	0.524	1.5	0.754	7.2	0.054
3.6	0.458	3.4	0.446	1.8	0.677	7.5	0.052
4.2	0.375	3.8	0.342	2.1	0.649	7.8	0.036
4.8	0.208	4.2	0.300	2.4	0.524	8.1	0.033
5.4	0.167	5.0	0.207	2.7	0.508	8.4	0.019
6.0	0.042	5.4	0.175	3.0	0.432	8.7	0.014
		5.8	0.148	3.3	0.389	9.3	0.012
		6.6	0.075	3.6	0.355	9.6	0.0069
		7.0	0.054	3.9	0.324	9.9	0.0062
		7.4	0.033	4.5	0.242	10.2	0.0027
		8.2	0.017	4.8	0.200	10.8	0.0016
		9.0	0.0017	5.1	0.190	11.1	0.00094
				5.4	0.158	12.0	0.000072

Table J   Critical values of the Siegel–Tukey statistic with sample sizes n and m $(n \leqslant m)$. One-tailed question: $\alpha = 0.05$

	n														
m	2	3	4	5	6	7	8	9	10	11	12	13	14	15
2														
3		6												
4	–	6	11											
5	3	7	12	19										
6	3	8	13	20	28									
7	3	8	14	21	29	39								
8	4	9	15	23	31	41	51							
9	4	10	16	24	33	43	54	66						
10	4	10	17	26	35	45	56	69	82					
11	4	11	18	27	37	47	59	72	86	100				
12	5	11	19	28	38	49	62	75	89	104	120			
13	5	12	20	30	40	52	64	78	92	108	125	142		
14	6	13	21	31	42	54	67	81	96	112	129	147	166	
15	6	13	22	33	44	56	69	84	99	116	133	152	171	192

Table K "Student's" t-test

The table gives the probability under H_0 of values of t as great as or greater than that obtained, given n d.f. (e.g., with $n = 6$, if we obtain a t of 3.0 we notice that its probability, given H_0 is between 0.05 and 0.02. We would thus reject H_0 if our α were 0.05, but not if it were 0.02)

n	Probability												
	0.9	0.8	0.7	0.6	0.5	0.4	0.3	0.2	0.1	0.05	0.02	0.01	0.001
1	0.158	0.325	0.510	0.727	1.000	1.376	1.963	3.078	6.314	12.706	31.821	63.657	636.619
2	0.142	0.289	0.445	0.617	0.816	1.061	1.386	1.886	2.920	4.303	6.965	9.925	31.598
3	0.137	0.277	0.424	0.584	0.765	0.978	1.250	1.638	2.353	3.182	4.541	5.841	12.924
4	0.134	0.271	0.414	0.569	0.741	0.941	1.190	1.533	2.132	2.776	3.747	4.604	8.610
5	0.132	0.267	0.408	0.559	0.727	0.920	1.156	1.476	2.015	2.571	3.365	4.032	6.869
6	0.131	0.265	0.404	0.553	0.718	0.906	1.134	1.440	1.943	2.447	3.143	3.707	5.959
7	0.130	0.263	0.402	0.549	0.711	0.896	1.119	1.415	1.895	2.365	2.998	3.499	5.408
8	0.130	0.262	0.399	0.546	0.706	0.889	1.108	1.397	1.860	2.306	2.896	3.355	5.041
9	0.129	0.261	0.398	0.543	0.703	0.883	1.100	1.383	1.833	2.262	2.821	3.250	4.781
10	0.129	0.260	0.397	0.542	0.700	0.879	1.093	1.372	1.812	2.228	2.764	3.169	4.537
11	0.129	0.260	0.396	0.540	0.697	0.876	1.088	1.363	1.796	2.201	2.718	3.106	4.437
12	0.128	0.259	0.395	0.539	0.695	0.873	1.083	1.356	1.782	2.179	2.681	3.055	4.318
13	0.128	0.259	0.394	0.538	0.694	0.870	1.079	1.350	1.771	2.160	2.650	3.012	4.221
14	0.128	0.258	0.393	0.537	0.692	0.868	1.076	1.345	1.761	2.145	2.624	2.977	4.140
15	0.128	0.258	0.393	0.536	0.691	0.866	1.074	1.341	1.753	2.131	2.602	2.947	4.076

Table K — *continued*

n	Probability												
	0.9	0.8	0.7	0.6	0.5	0.4	0.3	0.2	0.1	0.05	0.02	0.01	0.001
16	0.128	0.258	0.392	0.535	0.690	0.865	1.071	1.337	1.746	2.120	2.583	2.921	4.015
17	0.128	0.257	0.392	0.534	0.689	0.863	1.069	1.333	1.740	2.110	2.567	2.898	3.965
18	0.127	0.257	0.392	0.534	0.688	0.862	1.067	1.330	1.734	2.101	2.552	2.878	3.922
19	0.127	0.257	0.391	0.533	0.688	0.861	1.066	1.328	1.729	2.093	2.539	2.861	3.883
20	0.127	0.257	0.391	0.533	0.687	0.860	1.064	1.325	1.725	2.086	2.528	2.845	3.850
21	0.127	0.257	0.391	0.532	0.686	0.859	1.063	1.323	1.721	2.080	2.518	2.831	3.819
22	0.127	0.256	0.390	0.532	0.686	0.858	1.061	1.321	1.717	2.074	2.508	2.819	3.792
23	0.127	0.256	0.390	0.532	0.685	0.858	1.060	1.319	1.714	2.069	2.500	2.807	3.767
24	0.127	0.256	0.390	0.531	0.685	0.857	1.059	1.318	1.711	2.064	2.492	2.797	3.745
25	0.127	0.256	0.390	0.531	0.684	0.856	1.058	1.316	1.708	2.060	2.485	2.787	3.725
26	0.127	0.256	0.390	0.531	0.684	0.856	1.058	1.315	1.706	2.056	2.479	2.779	3.707
27	0.127	0.256	0.389	0.531	0.684	0.855	1.057	1.314	1.703	2.052	2.473	2.771	3.690
28	0.127	0.256	0.389	0.530	0.683	0.855	1.056	1.313	1.701	2.048	2.467	2.763	3.674
29	0.127	0.256	0.389	0.530	0.683	0.854	1.055	1.311	1.699	2.045	2.462	2.756	3.659
30	0.127	0.256	0.389	0.530	0.683	0.854	1.055	1.310	1.697	2.042	2.457	2.750	3.646
40	0.126	0.255	0.388	0.529	0.681	0.851	1.050	1.303	1.684	2.021	2.423	2.704	3.551
60	0.126	0.254	0.387	0.527	0.679	0.848	1.046	1.296	1.671	2.000	2.390	2.660	3.460
120	0.126	0.254	0.386	0.526	0.677	0.845	1.041	1.289	1.658	1.980	2.358	2.617	3.373
∞	0.126	0.253	0.385	0.524	0.674	0.842	1.036	1.282	1.645	1.960	2.326	2.576	3.291

Table L The F-ratio test

The obtained value of F enables us to reject H_0 for $\alpha = 0.05$ (upper row of each pair) or $\alpha = 0.01$ (lower row of each pair) if it is equal to or greater than that shown in the table

Degrees of freedom for greater mean square

Each cell shows the upper value ($\alpha = 0.05$) / lower value ($\alpha = 0.01$).

	1	2	3	4	5	6	7	8	9	10	11	12	14	16	20	24	30	40	50	75	100	200	500	∞
1	161 / 4052	200 / 4999	216 / 5403	225 / 5625	230 / 5764	234 / 5859	237 / 5928	239 / 5981	241 / 6022	242 / 6056	243 / 6082	244 / 6106	245 / 6142	246 / 6169	248 / 6208	249 / 6234	250 / 6258	251 / 6286	252 / 6302	253 / 6323	253 / 6334	254 / 6352	254 / 6361	254 / 6366
2	18.51 / 98.49	19.00 / 99.01	19.16 / 99.17	19.25 / 99.25	19.30 / 99.30	19.33 / 99.33	19.36 / 99.34	19.37 / 99.36	19.38 / 99.38	19.39 / 99.40	19.40 / 99.41	19.41 / 99.42	19.42 / 99.43	19.43 / 99.44	19.44 / 99.45	19.45 / 99.46	19.46 / 99.47	19.47 / 99.48	19.47 / 99.48	19.48 / 99.49	19.49 / 99.49	19.49 / 99.49	19.50 / 99.50	19.50 / 99.50
3	10.13 / 34.12	9.55 / 30.81	9.28 / 29.46	9.12 / 28.71	9.01 / 28.24	8.94 / 27.91	8.88 / 27.67	8.84 / 27.49	8.81 / 27.34	8.78 / 27.23	8.76 / 27.13	8.74 / 27.05	8.71 / 26.92	8.69 / 26.83	8.66 / 26.69	8.64 / 26.60	8.62 / 26.50	8.60 / 26.41	8.58 / 26.30	8.57 / 26.27	8.56 / 26.23	8.54 / 26.18	8.54 / 26.14	8.53 / 26.12
4	7.71 / 21.20	6.94 / 18.00	6.59 / 16.69	6.39 / 15.98	6.26 / 15.52	6.16 / 15.21	6.09 / 14.98	6.04 / 14.80	6.00 / 14.66	5.96 / 14.54	5.93 / 14.45	5.91 / 14.37	5.87 / 14.24	5.84 / 14.15	5.80 / 14.02	5.77 / 13.93	5.74 / 13.83	5.71 / 13.74	5.70 / 13.69	5.68 / 13.61	5.66 / 13.57	5.65 / 13.52	5.64 / 13.48	5.63 / 13.46
5	6.61 / 16.26	5.79 / 13.27	5.47 / 12.06	5.19 / 11.39	5.05 / 10.97	4.95 / 10.67	4.88 / 10.45	4.82 / 10.27	4.78 / 10.15	4.74 / 10.05	4.70 / 9.96	4.68 / 9.89	4.64 / 9.77	4.60 / 9.68	4.56 / 9.55	4.53 / 9.47	4.50 / 9.38	4.46 / 9.29	4.44 / 9.24	4.42 / 9.17	4.40 / 9.13	4.38 / 9.07	4.37 / 9.04	4.36 / 9.02
6	5.99 / 13.74	5.14 / 10.92	4.76 / 9.78	4.53 / 9.15	4.39 / 8.75	4.28 / 8.47	4.21 / 8.26	4.15 / 8.10	4.10 / 7.98	4.06 / 7.87	4.03 / 7.79	4.00 / 7.72	3.96 / 7.60	3.92 / 7.52	3.87 / 7.39	3.84 / 7.31	3.81 / 7.23	3.77 / 7.14	3.75 / 7.09	3.72 / 7.02	3.71 / 6.99	3.69 / 6.94	3.68 / 6.90	3.67 / 6.88
7	5.59 / 12.25	4.74 / 9.55	4.35 / 8.45	4.12 / 7.85	3.97 / 7.46	3.87 / 7.19	3.79 / 7.00	3.73 / 6.84	3.68 / 6.71	3.63 / 6.62	3.60 / 6.54	3.57 / 6.47	3.52 / 6.35	3.49 / 6.27	3.44 / 6.15	3.41 / 6.07	3.38 / 5.98	3.34 / 5.90	3.32 / 5.85	3.29 / 5.78	3.28 / 5.75	3.25 / 5.70	3.24 / 5.67	3.23 / 5.65
8	5.32 / 11.26	4.46 / 8.65	4.07 / 7.59	3.84 / 7.01	3.69 / 6.63	3.58 / 6.37	3.50 / 6.19	3.44 / 6.03	3.39 / 5.91	3.34 / 5.82	3.31 / 5.74	3.28 / 5.67	3.23 / 5.56	3.20 / 5.48	3.15 / 5.36	3.12 / 5.28	3.08 / 5.20	3.05 / 5.11	3.03 / 5.06	3.00 / 5.00	2.98 / 4.96	2.96 / 4.91	2.94 / 4.88	2.93 / 4.86
9	5.12 / 10.56	4.26 / 8.02	3.86 / 6.99	3.63 / 6.42	3.48 / 6.06	3.37 / 5.80	3.29 / 5.62	3.23 / 5.47	3.18 / 5.35	3.13 / 5.26	3.10 / 5.18	3.07 / 5.11	3.02 / 5.00	2.98 / 4.92	2.93 / 4.80	2.90 / 4.73	2.86 / 4.64	2.82 / 4.56	2.80 / 4.51	2.77 / 4.45	2.76 / 4.41	2.73 / 4.36	2.72 / 4.33	2.71 / 4.31
10	4.96 / 10.04	4.10 / 7.56	3.71 / 6.55	3.48 / 5.99	3.33 / 5.64	3.22 / 5.39	3.14 / 5.21	3.07 / 5.06	3.02 / 4.95	2.97 / 4.85	2.94 / 4.78	2.91 / 4.71	2.86 / 4.60	2.82 / 4.52	2.77 / 4.41	2.74 / 4.33	2.70 / 4.25	2.67 / 4.17	2.64 / 4.12	2.61 / 4.05	2.59 / 4.01	2.56 / 3.96	2.55 / 3.93	2.54 / 3.91
11	4.84 / 9.65	3.98 / 7.20	3.59 / 6.22	3.36 / 5.67	3.20 / 5.32	3.09 / 5.07	3.01 / 4.88	2.95 / 4.74	2.90 / 4.63	2.86 / 4.54	2.82 / 4.46	2.79 / 4.40	2.74 / 4.29	2.70 / 4.21	2.65 / 4.10	2.61 / 4.02	2.57 / 3.94	2.53 / 3.86	2.50 / 3.80	2.47 / 3.74	2.45 / 3.70	2.42 / 3.66	2.41 / 3.62	2.40 / 3.60
12	4.75 / 9.33	3.88 / 6.93	3.49 / 5.95	3.26 / 5.41	3.11 / 5.06	3.00 / 4.82	2.92 / 4.65	2.85 / 4.50	2.80 / 4.39	2.76 / 4.30	2.72 / 4.22	2.69 / 4.16	2.64 / 4.05	2.60 / 3.98	2.54 / 3.86	2.50 / 3.78	2.46 / 3.70	2.42 / 3.61	2.40 / 3.56	2.36 / 3.49	2.35 / 3.46	2.32 / 3.41	2.31 / 3.38	2.30 / 3.36

Table L — *continued*

Degrees of freedom for greater mean square

	1	2	3	4	5	6	7	8	9	10	11	12	14	16	20	24	30	40	50	75	100	200	500	∞
13	4.67 9.07	3.80 6.70	3.41 5.74	3.18 5.20	3.02 4.86	2.92 4.62	2.84 4.44	2.77 4.30	2.72 4.19	2.67 4.10	2.63 4.02	2.60 3.96	2.55 3.85	2.51 3.78	2.46 3.67	2.42 3.59	2.38 3.51	2.34 3.42	2.32 3.37	2.28 3.30	2.26 3.27	2.24 3.21	2.22 3.18	2.21 3.16
14	4.60 8.86	3.74 6.51	3.34 5.56	3.11 5.03	2.96 4.69	2.85 4.46	2.77 4.28	2.70 4.14	2.65 4.03	2.60 3.94	2.56 3.86	2.53 3.80	2.48 3.70	2.44 3.62	2.39 3.51	2.35 3.43	2.31 3.34	2.27 3.26	2.24 3.21	2.21 3.14	2.19 3.11	2.16 3.06	2.14 3.02	2.13 3.00
15	4.54 8.68	3.68 6.36	3.29 5.42	3.06 4.89	2.90 4.56	2.79 4.32	2.70 4.14	2.64 4.00	2.59 3.89	2.55 3.80	2.51 3.73	2.48 3.67	2.43 3.56	2.39 3.48	2.33 3.36	2.29 3.29	2.25 3.20	2.21 3.12	2.18 3.07	2.15 3.00	2.12 2.97	2.10 2.92	2.08 2.89	2.07 2.87
16	4.49 8.53	3.63 6.23	3.24 5.29	3.01 4.77	2.85 4.44	2.74 4.20	2.66 4.03	2.59 3.89	2.54 3.78	2.49 3.69	2.45 3.61	2.42 3.55	2.37 3.45	2.33 3.37	2.28 3.25	2.24 3.18	2.20 3.10	2.16 3.01	2.13 2.96	2.09 2.89	2.07 2.86	2.04 2.80	2.02 2.77	2.01 2.75
17	4.45 8.40	3.59 6.11	3.20 5.18	2.96 4.67	2.81 4.34	2.70 4.10	2.62 3.93	2.55 3.79	2.50 3.68	2.45 3.59	2.41 3.52	2.38 3.45	2.33 3.35	2.29 3.27	2.23 3.16	2.19 3.08	2.15 3.00	2.11 2.92	2.08 2.86	2.04 2.79	2.02 2.76	1.99 2.70	1.97 2.67	1.96 2.65
18	4.41 8.28	3.55 6.01	3.16 5.09	2.93 4.58	2.77 4.25	2.66 4.01	2.58 3.85	2.51 3.71	2.46 3.60	2.41 3.51	2.37 3.44	2.34 3.37	2.29 3.27	2.25 3.19	2.19 3.07	2.15 3.00	2.11 2.91	2.07 2.83	2.04 2.78	2.00 2.71	1.98 2.68	1.95 2.62	1.93 2.59	1.92 2.57
19	4.38 8.18	3.52 5.93	3.13 5.01	2.90 4.50	2.74 4.17	2.63 3.94	2.55 3.77	2.48 3.63	2.43 3.52	2.38 3.43	2.34 3.36	2.31 3.30	2.26 3.19	2.21 3.12	2.15 3.00	2.11 2.92	2.07 2.84	2.02 2.76	2.00 2.70	1.96 2.63	1.94 2.60	1.91 2.54	1.90 2.51	1.88 2.49
20	4.35 8.10	3.49 5.85	3.10 4.94	2.87 4.43	2.71 4.10	2.60 3.87	2.52 3.71	2.45 3.56	2.40 3.45	2.35 3.37	2.31 3.30	2.28 3.23	2.23 3.13	2.18 3.05	2.12 2.94	2.08 2.86	2.04 2.77	1.99 2.69	1.96 2.63	1.92 2.56	1.90 2.53	1.87 2.47	1.85 2.44	1.84 2.42
21	4.32 8.02	3.47 5.78	3.07 4.87	2.84 4.37	2.68 4.04	2.57 3.81	2.49 3.65	2.42 3.51	2.37 3.40	2.32 3.31	2.28 3.24	2.25 3.17	2.20 3.07	2.15 2.99	2.09 2.88	2.05 2.80	2.00 2.72	1.96 2.63	1.93 2.58	1.89 2.51	1.87 2.47	1.84 2.42	1.82 2.38	1.81 2.36
22	4.30 7.94	3.44 5.72	3.05 4.82	2.82 4.31	2.66 3.99	2.55 3.76	2.47 3.59	2.40 3.45	2.35 3.35	2.30 3.26	2.26 3.18	2.23 3.12	2.18 3.02	2.13 2.94	2.07 2.83	2.03 2.75	1.98 2.67	1.93 2.58	1.91 2.53	1.87 2.46	1.84 2.42	1.81 2.37	1.80 2.33	1.78 2.31
23	4.28 7.88	3.42 5.66	3.03 4.76	2.80 4.26	2.64 3.94	2.53 3.71	2.45 3.54	2.38 3.41	2.32 3.30	2.28 3.21	2.24 3.14	2.20 3.07	2.14 2.97	2.10 2.89	2.04 2.78	2.00 2.70	1.96 2.62	1.91 2.53	1.88 2.48	1.84 2.41	1.82 2.37	1.79 2.32	1.77 2.28	1.76 2.26
24	4.26 7.82	3.40 5.61	3.01 4.72	2.78 4.22	2.62 3.90	2.51 3.67	2.43 3.50	2.36 3.36	2.30 3.25	2.26 3.17	2.22 3.09	2.18 3.03	2.13 2.93	2.09 2.85	2.02 2.74	1.98 2.66	1.94 2.58	1.89 2.49	1.86 2.44	1.82 2.36	1.80 2.33	1.76 2.27	1.74 2.23	1.73 2.21
25	4.24 7.77	3.38 5.57	2.99 4.68	2.76 4.18	2.60 3.86	2.49 3.63	2.41 3.46	2.34 3.32	2.28 3.21	2.24 3.13	2.20 3.05	2.16 2.99	2.11 2.89	2.06 2.81	2.00 2.70	1.96 2.62	1.92 2.54	1.87 2.45	1.84 2.40	1.80 2.32	1.77 2.29	1.74 2.23	1.72 2.19	1.71 2.17

Table M   Spearman's ρ test

The table gives critical values of Spearman's correlation coefficient ρ

N	α 0.05	(One-tailed) 0.01	N	α 0.05	(One-tailed) 0.01
4	1.000		16	0.425	0.601
5	0.900	1.000	18	0.399	0.564
6	0.829	0.943	20	0.377	0.534
7	0.714	0.893	22	0.359	0.508
8	0.643	0.833	24	0.343	0.485
9	0.600	0.783	26	0.329	0.465
10	0.564	0.746	28	0.317	0.448
12	0.506	0.712	30	0.306	0.432
14	0.456	0.645			

Table N   Kendall's τ-test

The table gives the probabilities associated with values as large as obtained values of Kendall's correlation coefficient τ

	Values of N					Values of N		
τ	4	5	8	9	τ	6	7	10
0	0.625	0.592	0.548	0.540	1	0.500	0.500	0.500
2	0.375	0.408	0.452	0.460	3	0.360	0.386	0.431
4	0.167	0.242	0.360	0.381	5	0.235	0.281	0.364
6	0.042	0.117	0.274	0.306	7	0.136	0.191	0.300
8		0.042	0.199	0.238	9	0.068	0.119	0.242
10		0.0083	0.138	0.179	11	0.028	0.068	0.190
12			0.089	0.130	13	0.0083	0.035	0.146
14			0.054	0.090	15	0.0014	0.015	0.108
16			0.031	0.060	17		0.0054	0.078
18			0.016	0.038	19		0.0014	0.054
20			0.0071	0.022	21		0.00020	0.036
22			0.0028	0.012	23			0.023
24			0.00087	0.0063	25			0.014
26			0.00019	0.0029	27			0.0083
28			0.000025	0.0012	29			0.0046
30				0.00043	31			0.0023
32				0.00012	33			0.0011
34				0.000025	35			0.00047
36				0.0000028	37			0.00018
					39			0.000058
					41			0.000015
					43			0.0000028
					45			0.00000028

Table O   Kendall's *W*-test

The table gives critical values of Kendall's coefficient of concordance *W*

	N					*Additional values for N = 3*	
k	3†	4	5	6	7	*k*	*s*
Values for α = 0.05							
3			64.4	103.9	157.3	9	54.0
4		49.5	88.4	143.3	217.0	12	71.9
5		62.6	112.3	182.4	276.2	14	83.8
6		75.7	136.1	231.4	335.2	16	95.8
8	48.1	101.7	183.7	299.0	453.1	18	107.7
10	60.0	127.8	231.2	376.7	571.0		
15	89.8	192.9	349.8	570.5	864.9		
20	119.7	258.0	468.5	764.4	1158.7		
Values for α = 0.01							
3			75.6	122.8	185.6	9	75.9
4		61.4	109.3	176.2	265.0	12	103.5
5		80.5	142.8	229.4	343.8	14	121.9
6		99.5	176.1	282.4	422.6	16	140.2
8	66.8	137.4	242.7	388.3	579.9	18	158.6
10	85.1	175.3	309.1	494.0	737.0		
15	131.0	269.8	475.2	758.2	1129.5		
20	177.0	364.2	641.2	1022.2	1521.9		

Index

Only first and main accounts are given. The latter are in **bold type** when different from the former.